THE
SPIRITUAL
SCIENCE
OF THE
STARS

THE
SPIRITUAL
SCIENCE
OF THE
STARS

A GUIDE TO THE ARCHITECTURE OF THE SPIRIT

PETE STEWART

Inner Traditions
Rochester, Vermont

Inner Traditions
One Park Street
Rochester, Vermont 05767
www.InnerTraditions.com

Library of Congress Cataloging-in-Publication Data
Stewart, Pete.
  The spiritual science of the stars : a guide to the architecture of the spirit / Pete Stewart.
    p. cm.
  Summary: "The profound influence of ancient cosmologies on our ideas about the
human spirit"—Provided by publisher.
  Includes bibliographical references (p.   ) and index.
  ISBN-13: 978-1-59477-196-5
  ISBN-10: 1-59477-196-0
  1. Cosmology, Ancient—Miscellanea. 2. Civilization—History. 3. Knowledge, Theory
of (Religion) I. Title.
  BF1999.S755 2007
  202'.4—dc22
                                                                    2007032953

Printed and bound in the United States by Lake Book Manufacturing

10 9 8 7 6 5 4 3 2 1

Text design by Priscilla Baker and layout by Jon Desautels
This book was typeset in Garamond Premier Pro with Schneidler used as a display
typeface

To send correspondence to the author of this book, mail a first-class letter to the author
c/o Inner Traditions • Bear & Company, One Park Street, Rochester, VT 05767, and
we will forward the communication.

For Joel and Leo

CONTENTS

INTERLUDE

THE MYTH OF REUNION

ACKNOWLEDGMENTS

This work has been in progress in some form or another for more than thirty years. During that time, far too many people to acknowledge here have enjoyed or endured peacefully my obsessions and fancies. For their varying degrees of bewilderment, tolerance, and support, I am grateful. In seeing the final emergence of the book, I have to thank Richard Heath for the original introduction to Skyglobe software, without which much of the work could not have been done; Gordon Strachan, who first encouraged me to believe that the work was worthwhile; and Anthony Blake, who reminded me of that when I was in danger of forgetting it.

My biggest debt, however, remains to the late Hertha von Dechend and Giorgio de Santillana, whose book *Hamlet's Mill* first inspired me. If anything of lasting value lodges in these pages, it is due to their courage in expounding within academic circles ideas so radical that they could meet only with dismissal, such being the way of academia. I owe a similar debt to William Sullivan, whose work *The Secret of the Incas* arrived at a point where I had allowed myself to believe that academia was right and that such ideas were of little worth. His application of detailed scholarship to a much misunderstood topic was then, and remains still, a revelation. Working within areas such as this, which seem to challenge everything we think we know about our history and our sense of meaning, frequently leads to debilitating self-doubt; I have returned at frequent intervals to both these texts for reassurance that there is something here that can be transforming.

Research work in these areas necessitates access to wide-ranging sources. When I began in the 1970s, I was helped immensely by the extensive collection of obscure works donated by some earlier obsessive to the reference department of Sheffield City Library; it seems unlikely that the same staff helped me when I revisited these sources recently, but my belated thanks to them all wherever they are. In recent years the Internet has made such work a very different process; of the numerous sites I have found valuable, that at www.sacred-texts.com deserves special mention. It is, as it says on the home page, "A quiet place in cyberspace . . . Open Source for the Human Soul."

LIST OF
ILLUSTRATIONS

Except where noted, all drawings and diagrams are the work of the author.

Fig. 1. The Binding Serpent. Detail from an inscribed stone from Sjonhem Parish Church, Gotland, Sweden; County Museum of Gotland. Adapted from an image at www.gotmus.i.se.

Fig. 2. The Coils of Time. Vishnu asleep on the coils of the serpent Sesa. Detail from *Atha Naradiyamahapuranam* (Narada's Great Ancient Tale), Mumbayyam, India: Sri Venkatesvara Stim-Yantragare, 1923.

Fig. 3. Izanami and Izanagi Stirring the Ocean of Chaos. Detail from Tominobu Hosoda, *Kamiyo no masagoto tokiwagusa* (Mythological Story of the Creation of Japan), Kyoto, 1827.

Fig. 4. The Raised-up-Sky. Drawing of the cross at the Temple of the Cross, Palenque, by Linda Schele. © David Schele, courtesy Foundation for the Advancement of Mesoamerican Studies, Inc. (www.famsi.org).

Fig. 5. The Divided Sky

Fig. 6. The Royal Surveyors. Detail from a stone engraving in the tomb of the U family in the province of Shangtung, China, second century CE.

Introduction
MAKING SENSE

In the beginning of things, there was a time when the
sky was not very high up above the heads of men. It was
no higher than the top of the tent in which they lived.
Then it was easy for people to communicate with the
celestial deities through the opening left in the top of the
tent. But according to certain legends, one day a woman
complained that smoke had got into the tent unnecessarily
and this angered the spirits, who then sent a giant to
lift the sky and raise it to its present height. From then
on men found it necessary to have sorcerers to intercede
between themselves and the celestial deities.

THE ORIGINAL STORY

The story of how heaven and Earth came to be divided, of how people
were separated from their gods and set out on their journey through
time, must be one of the most fundamental of all stories. The version
at the head of this chapter is from the Samoyed Yuraks of the far north
of Europe. In the West, we are perhaps more likely to be familiar with
the version from classical Greece in which Kronos cuts apart his parents, Sky and Earth (Ouranos and Gaia), putting an end to their primordial embrace. But these are only two of the multitude of ways of

telling a story that is known in some form or other almost everywhere in the world.

Sometimes the story has the air of a folktale, as in the version from several continents that tells how the women, pounding grain with their pestles, banged against the body of Heaven, angering him and causing him to raise himself above the Earth,[1] or that from the Banaba tradition in the Pacific islands, where Auriaria, the lord of all lands and a mighty giant, made a doorway through the rock of heaven and entered beneath, where he pushed upon heaven from the underside and raised it a great height from earth.[2] Sometimes it appears within a story about something very different, such as that from the Australian Aboriginal people of the Murrumbidgee River, which tells how a chief, while searching for his runaway wife, pulls a golden rod from the waters of a lake and finds that he has pushed up the sky, which previously had been close down on the earth.[3]

At other times the story is cloaked in all the awesome mystery of the most ancient scriptures, as in the Hindu hymn to Varuna, the creator and guardian of the sacred law, who is said to have "propped apart the two world-halves even though they are so vast. He has pushed away the dome of the sky to make it high and wide."[4]

To judge by both its variety and its universality, this is clearly a story that has had profound importance for tellers and listeners alike. The ideas it expresses seem to have played a major role in determining human attitudes for thousands of years. In the region of Central America occupied by the Mayan people, to this day an annual ceremony is reenacted wherein the two priests and their wives, having previously set out the four foundation stones of the "earth-sky," rise with great precision from their seats at the four corners; they are said to be "lifting the sky."[5]

The essence of our "original" story is this: the world that was created in the beginning was one of timeless accord, characterized by a unique and harmonious relationship between heaven and earth, when gods and men communed. Adversity of any kind—whether war, sickness, or death—was unknown. As the result of some misdemeanor on the part of the people, which involved them in some way overstepping their mark, this harmonious relationship was disrupted and the perfect world came to an end. At this point, time began, with all its consequences. Not only the people but also the world itself was destined for decay, death, and disaster.*

The human response to the situation described in this story varies a good deal in emphasis, depending on cultural conditions. Generally, however, it has been to look forward to a time when this perfect world can be reattained, whether that be at the end of an individual life or at the end of the life of the world itself. There have even been those who, with varying motives, have envisaged this perfect world being established right here on earth.

All these responses can be summed up in what Mircea Eliade called "the myth of re-integration." He writes: "Almost everywhere in the history of religion . . . in an infinity of variations . . . fundamentally it is an expression of the thirst to abolish dualism, endless returnings and fragmentary existences. It existed at the most primitive stages, which indicates that man, from the time when he first realised his position in the universe, desired passionately, and tried to achieve concretely, a passing beyond his human status."[6]

When we realize that the description of the events that took place in Eden, when Adam and Eve were driven from the company of their God in paradise, is itself a version of this story, then we can see that this is truly the story of the beginning of human thought as we know it now. The variety of forms in which it is told may be vast, but the significance is unvarying. The events it describes form the heart of every attempt to perceive meaning in the human situation. Whatever sense has been made of human experience turns out to be built on the ground plan this story establishes.

For thousands of years, the meaning with which these stories have

---

*It might be argued that there are two stories here, one describing the lifting of the sky at the first creation of the world and a second describing the separation at the end of that creation. It will become clear how these two interpretations are linked inseparably.

informed our existence has been derived from the assumption that they are descriptions of actual events. It is hardly surprising, therefore, that our growing questioning of the factual validity of this story over the last century or so has led to a gradual collapse of that meaning, with all the destabilizing consequences that implies.

During the last twenty or thirty years, however, as this source of meaning was beginning to look as if it were about to dry up completely, leaving humanity alone and facing a disinterested universe, an extraordinary change in understanding has been taking place. As science has been putting the finishing touches to its own story of creation, a story that seems to put chance in the place of purpose, the old stories have begun to reveal a very different significance. From behind their apparently straightforward descriptions, there has begun to emerge another, more profound and enduring meaning.

As long as these stories were believed unquestioningly, such a meaning would have gone unnoticed. But as we question them more, it begins to seem possible that, rather than reject them as fantasies generated by human minds in their infancy, or broaden their meaning to accommodate every new turn of cosmological development, we might begin to see them in a quite different light. Suddenly it seems that the whole significance of the myth of creation may have been misunderstood, that it may be possible to understand it in a wholly new way, a way that restores to it just that profundity and precision of which contemporary cosmology has deprived it.

This book attempts to outline what such a reinterpretation might look like. What emerges is shattering to our sense both of contemporary meaning and of ancient prehistory. It does, however, offer a suggestion as to how new senses of meaning might evolve from our rapidly changing understanding of the universe.

The book is divided into three major sections, each of which deals with one of the three determining myths that describe the nature of creation: first, the myth of creation itself; second, what I have called the myth of corruption, which tells how it became necessary to divide heaven from earth, and how, as a result, humanity was divided from its creators; and lastly, the myth of reunion, which tells how the world might be regenerated and the separated elements reconciled, creation and humanity sharing in an immortal and eternal existence.

Among them, these three stories tell of the very beginning of human speculation. The variety of forms in which they are told may be vast, but the significance is unvarying. They form the heart of every subsequent attempt to perceive meaning in the human situation. To ask how they came to be told is to ask how the human mind came to be the way it is.

The first tellers of these stories clearly read whatever formative events inspired them as messages concerning the meaning of the world around them. In telling these experiences as stories for the first time, they allowed the human imagination to begin to construct a ground plan upon which all its future speculations might stand. They began, in fact, to make sense.

Stories do not tell themselves, however. There must have been a time, remote in human evolution as it might be, before which these stories and their consequences for human aspiration did not exist. Although they may appear to form a sequence, these stories are interdependent. They make sense only as a whole.

We shall see that crucial to the stories' whole structure is the central idea of severance. Not only does this concept generate the need for reconciliation, but in many instances, in some strange way, it is also the cause of creation itself. Seen in this light, there is no way in which the stories could be interpreted as history, even if we could accept as credible a literal occurrence of the events many of them describe.

These stories have shaped the way we view the world and the sense we make of our experience of it. So we have to ask what kind of experience could have first caused mankind to "realize his position in the universe" and to describe it in this way. What kind of inspiration can have been so profound that its effects would echo through millennia, resulting in that universal thirst for reintegration that has been so powerful and enduring in its shaping of the human spirit?

REALMS UNCHARTED

... hieroglyphs old
which sages and keen eyed astrologers
Then living on earth, with labouring thought
Won from the gaze of many centuries:
Now lost, save what we find on remnants huge
Of stone, or marble swart, their importance gone
Their wisdom long since fled

KEATS, *HYPERION*, BOOK I, 274–83

Around thirty years ago, the experts were in agreement that the prehistory of human culture had been fairly comprehensively mapped. There might be some controversies about dating to be resolved, but the basic plan had been firmly established. The rungs on the ladder of technological development were firmly in place. From those remote days of the early hominids in Africa, human consciousness had gradually emerged through the well-known levels of the "Stone Age," from the pre-glacial Paleolithic to the Mesolithic and on toward the flowering of writing and history itself in the early Neolithic. It was only with this latest stage that any kind of thought process evolved that could take consciousness beyond the "primitive" stage of awe-inspired reaction to the uncontrollable forces of nature. It seemed there was little more but detailed infill to learn.

And then something very remarkable began to happen. A few dissenting voices of scholarship began to point to totally inexplicable evidence. Was it possible, the suggestion went, that perhaps there were things about which we really knew very little? Was what we did know based almost entirely on misunderstanding, and sometimes simply ignoring, such evidence? The mainstream of scholarship in prehistory and archaeology was aghast and refused to consider these developments. To accept them might be to upend the whole structure that had been so laboriously constructed. Put simply, the proposal that caused, and still causes, such controversy and indignation was that, from at least as far back as the end of the last ice age, and perhaps for the entire history of *Homo sapiens sapiens,* one factor, hitherto almost totally ignored, had dominated the emergence of human culture. Above all else, we might say, our ancestors were obsessed with the study of the night sky.

Scholars of ancient history had long appreciated the formative role that "astrology" had played in the flowering of Mesopotamian culture, but what was now being suggested went much further. Far from emerging out of the city-states of Assyria and Babylon, this obsession with astronomical events was shown to be almost universal and showed evidence of being far more ancient. It was the emergence of the new discipline of archaeoastronomy that first introduced this heresy, surveying and measuring the astronomical alignments identified as being built into megalithic sites. But few were prepared to say what motivated the immense labor that went into this obsessive observation. Why were these "primitive" people so concerned to record the circlings of the stars?

An answer of a kind was actually beginning to emerge at around the same time, in the form of a dramatic reinterpretation of the material of mythology. This material, both from ancient documents and from contemporary anthropology, began to be seen as adding a kind of "text" to the ancient megalithic structures. These stories can be heard to talk insistently a stellar language.

One of the first works to outline the grammar of this stellar language was *Hamlet's Mill*,[7] a groundbreaking book first published in 1969. Reading the story of how this work came to be written, one gets the impression that here were ancient ideas that had lain dormant and cloaked in almost willfully misleading disguises for millennia, finally deciding that their time had come.

It was a state of bewilderment at the vast amount of unintelligible material facing her that first led the historian of science Hertha von Dechend to uncover its secret ("To call it being struck by lightning would be more correct," she says). She had been studying Polynesian myth in a search for an understanding of the *"deus faber,"* the creator/craftsman god found in almost all cultures, but had come to the realization that she really understood nothing of the thousands of pages of material she had read. "The annihilating recognition of our complete ignorance came down on me like a sledge hammer," she writes. "There was no single sentence that could be understood."[8]

And yet the Polynesians were capable of navigating their tiny craft across the world's greatest ocean; there seemed little reason not to take their intellect seriously. Von Dechend explains how, while studying the archaeological remains on the many small islands, "a clue was given to me which I duly followed up." (She had discovered that, of the two of these islands most densely covered with a particular kind of cult place or temple, one was located on the Tropic of Capricorn and the other on the Tropic of Cancer.) "Having come to the history of science through the study of ethnology . . . there existed 'in the beginning' only the firm decision never to become involved in astronomical matters, under any condition," she observes.[9]

There had been attempts in the past to show that at least some, if not all, of mythology might in some way have been derived from the observation of events in the night sky. The suggestion was not uncommon in the nineteenth century. Much evidence was produced purporting to prove the existence of such observation and the importance attached to it, but no really consistent theory was ever proposed to make such an interpretation convincing. Attempts to establish a relationship between myth and astronomy fell into disrepute and total neglect as the interpretation of myth in terms of Jungian psychology took center field.

Although von Dechend had felt that Plato might provide a better source of insight into the essence of myth than psychologists could, and had experienced a "growing wrath" about the current interpretations, she had "least of all" the intention to explore the astronomical nature of myth. Until, that is, the revelation of the significance of the locations of those Polynesian islands. "And then there was no salvation any more," she declares; "astronomy could not be escaped."[10]

The culmination of Hertha von Dechend's inspired insight was the publication, together with fellow historian of science Giorgio de Santillana, of their extraordinary work *Hamlet's Mill*. Santillana describes this vast book as "only an essay . . . a first reconnaissance of a realm well-nigh unexplored and uncharted."

In *The Secret of the Incas,* William Sullivan describes his experience of first reading this work, an experience that anyone who has encountered it will recognize: "The ideas in *Hamlet's Mill* staggered me. The book stood conventional notions of 'prehistory' on their head. . . . The implications of *Hamlet's Mill* appeared to me nothing short of revolutionary."[11]

When I first read this "essay" in 1979 (in fact, the book is nearly five hundred pages, and its argument follows a tumultuous, not to say unruly, course), I felt that a window had been opened up across a vista that had once seemed familiar and which now could never look the same again. Once there had been stories that, although they seemed to make no acceptable sense except as fanciful parables, were called on to support the most profound of thoughts. Now, from the vantage point of this newly opened window, it was possible to see a realm where these stories not only made sense but also formed the most solid foundation. It was the later thinking, the constructs of philosophy and religion, that began to take on the air of parable.

THE LANGUAGE OF THE STARS

Although the realm of which *Hamlet's Mill* is a "reconnaissance" is vast and often impenetrable, the view we are offered across it originates in one quite simple idea, and one that is not new. Even Freud was aware that "man's observations of the great astronomical periodicities not only furnished him with a model, but formed the ground plan of his first attempts to introduce order into his life."

We have grown so used to the vague reddish glow of our urban night skies that it is sometimes difficult to reconstruct the overpowering effect that a clear, starry heaven can have. At some early stage in the emergence of the idea of ordering experience (the beginnings of self-consciousness), when the night sky was everywhere unpolluted, one poet, or perhaps many poets, saw the starry sky as an inspiration concerning the form of that ordering.

These "exceptional men," as Santillana calls them, unerringly read the vault of heaven as a message in which were encoded the laws of creation. Watching the progress of the heavenly lights night by night, year by year, they must have believed they were observing the pattern according to which their lives and those of all others, animate and inanimate, were governed, that they were watching the broadcast of the most majestic model possible of the workings of the universe.*

In his book *Echoes of the Ancient Skies,* archaeoastronomer E. C. Krupp says:

> For most of the history of humankind, going back to stone age times, the sky has served as a tool. . . . And just as their culture was partly a product of the tools they made . . . it was also shaped by their perception of the sky. From the sky they gained—and we, their descendants, have inherited—a profound sense of cyclic time, of order and symmetry, and of the predictability of nature. In this awareness lie not only the foundation of science but of our view of the universe and our place in it.[12]

Here, then, was the reason why our ancestors fixed their gaze so fervently on the heavens. "The Secret of Being lay displayed before their eyes," says Santillana. This secret turns out to be the ultimate source from which the entire fabric of meaning has been woven.

The idea that, from the very earliest times, the night sky was being read as a message concerning the laws of organization becomes in Santillana and von Dechend's book the basis for a total reappraisal, not only of the meaning of myth but of the whole idea of preliterate history as well. This was possible as a consequence of two realizations. The first was that the "astronomical periodicities," so often taken as being the obvious ones of day and night, winter and summer, should be expanded to include all periodicities discernible from the earth, including the most vast and elusive ones. This was perhaps the most radical

---

*In the introductory chapter to his book *Road to Reality* (London: Knopf, 2005, 1ff), Roger Penrose includes a fanciful narrative that records this experience, in which the stars are seen to remain undisturbed after cataclysmic earthly events.

aspect of their work. To accept it meant to rewrite all the prehistory of understanding, since it implied that this knowledge was of far greater antiquity than had been assumed.

The second realization was more subtle, and its consequences have barely begun to be grasped. It involved the identification of the fundamental clue to interpreting the material of myth. Essential to this was the recognition of the general tendency of myth to use everyday language to describe notions that are far from everyday.

Whatever the original purpose of this kind of encodement, what von Dechend and Santillana offered were the essential "translations." These everyday terms became understandable in their cosmological meanings. Thus we have more or less precise cosmological definitions for familiar words such as *earth* and *heaven, fire,* and *water.* With the help of these insights, meaningless nonsense and fancy are transformed into profound statements about the order of the world.

It would be hard to overestimate the impact of the revelations contained in *Hamlet's Mill* on the task of understanding myth. Not only did it propose to revolutionize notions of preliterate learning, but it also gave us the necessary groundwork upon which to construct a complete reassessment of mythical material of all kinds. In essence, the ancient code by means of which knowledge of the processes of the heavens was metaphorically encrypted within myth has begun to be broken. We can start to translate our familiar tales from myth into the language of ancient astronomy.

Perhaps the most challenging element in this task is to realize that, in addition to the terms identified by *Hamlet's Mill* as having cosmological meanings, we must now consider both *world* and *creation* themselves as being technical terms with their own precise astronomical meanings. Even to make such a suggestion seems to fly in the face of common sense. Once the possibility is allowed, however, a much grander sense can emerge, one that is far from common. The result is a revelation of the process whereby the secret of being was spun out into the fabric of meaning.

In his book *Consciousness Explained,* Daniel C. Dennett describes what he calls "the function of brains"; their purpose, he says, is "to produce future. . . . Faced with the task of extracting useful future out of our personal pasts, we organisms try . . . to find the laws of the world—and if there aren't any, to find approximate laws."[13]

The ancient observers, driven by this primal urge, read the starry cycles as the ultimate source of such laws. They diligently watched the patterns of creation evolve before them. In accordance with these patterns, spinning the thread of metaphor, they began to weave the extraordinary texture of meaning.

So profound was the effect of the recognition of these patterns that they formed the basis of the structure that ancient observers built out of their insights into the operations of the universe. Thanks to the pioneering work of Santillana and von Dechend, it is now possible to draw up a working plan of this structure. Within the measures of this plan, we can recognize the origins of our own most cherished ideas about meaning.

To appreciate the profundity of such a structure is hardly likely to be simple. Our quest involves the exploration of the grandest cycles in the visible cosmos. It will require the attitude of the explorer, prepared to venture boldly into unknown realms. For the most part, we shall be traveling in country "well-nigh uncharted," and we shall have to write our own guidebook. Even the few recognizable landmarks we encounter may turn out to point us in unfamiliar directions. As we travel farther, however, what seemed at the beginning to be the most obscure territory will become increasingly familiar.

There are in truth very few characters in the story, although they may come in many disguises and speak in uncouth tongues. Their initially bewildering appearance and behavior and the strangeness of the environments they inhabit will gradually take on the shape of a plot. The denouement of this plot is nothing less than revolutionary for our ideas about meaning. From the kind of material that we normally take to have little or no relevance to the world we live in we shall piece together layer by layer a most extraordinary structure. This structure, which has formed the basic framework for the whole development of human understanding, truly deserves Lévi-Strauss's description, "The Architecture of the Spirit."

The Myth of Creation

OUT OF THE ABYSS

Before the seas, and this terrestrial ball, the World
And Heav'n's high canopy, that covers all,
One was the face of Nature, if a face:
Rather a rude and indigested mass:
A lifeless lump, unfashion'd, and unfram'd,
Of jarring seeds; and justly Chaos nam'd.

OVID, *METAMORPHOSES*, BOOK I

THE YAWNING GAP

What existed before existence began? Even after thousands of years of seeking answers, simply asking such a question seems to tax the human imagination to the breaking point. Since 1929, however, the question has had new significance. In that year, the astronomer Edwin Hubble made a series of observations that were to lead to one of the greatest intellectual challenges in the history of natural philosophy. He showed that distant galaxies are moving away from us at high speeds and that this is true regardless of which direction we look in. This fact was ultimately to shatter the orthodox view of creation, bringing the idea from the realm of religion firmly into the realm of science, for the conclusion drawn from Hubble's research was that there had once been a time when all the matter in the universe had been gathered together in an infinitely dense mass occupying an infinitely small volume.

14

The universe could thus be shown to have had a discernible begin-ning, or at least something that might be considered to be a beginning. Nevertheless, this knowledge could not provide an answer to our ini-tial question. Stephen Hawking explains: "Because mathematics cannot really handle infinite numbers . . . there is a point in the universe where the [general theory of relativity] itself breaks down. Such a point is an example of what mathematicians call a singularity. . . . All our theories of science . . . break down at the big bang singularity, where the curva-ture of space-time is infinite. . . . We should therefore say that Time had a beginning at the Big Bang."[1]

That is to say, we cannot speak of a "before" in relation to the begin-ning. Even if there were events outside of this "singularity," Hawking goes on to say, no information could come to us from there. Hawking quotes St. Augustine, who had long since made this point, saying that time was a property of the universe that God created, and that time did not exist before the beginning of the universe.

We shall come to see during the course of this book that it might be more helpful to consider "world" to be a property of the time that God created. Theologians, faced with the expanding universe, have embarked on a process of endless revision of the religious interpretation of the creation story. Such a process is really necessary only if we insist on regarding myth as history, albeit very ancient history. This kind of imprisonment within our own particular time-bound logic of cause and effect can only return us continually to confront our first unanswerable question.

The myths, in contrast, are happy not only to gaze out across the yawning gap of preexistence, but also to report on what is to be seen there. They operate according to their own logic, wherein the sole pur-pose of "before" is not to cause but to explain "after," to stretch out a blank canvas on which a map of creation can be drawn. Such a "before" need contain only an absence of what is to come. Hence the opening of the Rig Veda hymn to creation: "There was neither non-existence nor existence then."[2]

This may not amount to a particularly satisfying answer. There is more, however. This "neither/nor" can be described. The description given in third-dynasty Egypt is typical: "In the beginning there was nothing but immense chaos, the deity Nun, thought to be an ocean or shapeless magma."[3]

The primary role of this description of preexistence, and it is an almost universal one—whether it appears as an ocean, an abyss, or more generally "the Waters" or "the Deep"—is everywhere to encompass the notion of formlessness. The Book of Genesis is most explicit about it. At the beginning, when God was creating the world, "the earth was without form; and darkness was on the face of the deep."[4]

Genesis here offers us a further characteristic of nonexistence, which is common to most traditions: the absence of light. The Maori cosmology succinctly combines these two features of the state that prevailed before the creation: "The universe was in darkness, with water everywhere."[5]

The insistence on formlessness and darkness may at first seem to be only the inevitable consequence of attempting to cross beyond the bounds of language and describe the unknowable. But the lack of shape in preexistence is in fact the clue to the nature of the whole process of creation. Just as the expansion of the universe described in the big bang theory allows time and space to enter the universe, thereby affording the means by which it is possible to measure its development, so mythology describes creation as the process whereby definition is imposed upon the boundless, enabling the proper order of things to emerge.

We would be making a serious error, however, if we were to allow the apparent resemblance between ancient and contemporary notions about the beginning of the universe in time to mislead us into thinking that the theories of St. Augustine and Stephen Hawking concern themselves with the same events. The Hindu myth, for instance, tells us, "In the beginning the universe was shrouded, indiscernible, undiscoverable, unknowable."[6]

At first sight this appears to be in agreement with Hawking and Aquinas in holding that no information can come to us from beyond the beginning. Hawking, however, has identified two branches of

physics, one that seeks to uncover the state of the universe at its beginning and another, more familiar perhaps, that seeks to comprehend
the laws that govern the behavior of the perceivable world.[7] Despite
the apparent concern of the cosmological myths with the former, they
are in fact concerned only with the latter. They deal not with the
beginning of the universe as such but with the beginning of *cosmos,*
with what the myths refer to as the creation, but that we might better call the identification, of order within the universe. The "unknowable" universe is actually awaiting the process of discernment so that
it might yield up its information.

BOUNDARYLESS EVERYTHING

The description of the creation at the opening of the Book of Genesis
offers further insight into this process when we consider the Hebrew
text, which makes reference to the *tohu* and *bohu,* translated in the King
James Bible as "formless void." The Jerusalem Bible commentary tells us
that this passage implies that "in the beginning," the earth was a "trackless waste." It is the kind of text that we too easily pass over as mere metaphor. The absence of any clear pathway across this waste is really quite
precise and revealing. For the Greeks, this was the "pathless ocean." To
voyage upon it was considered to be meaningless, since no point of reference existed within it to distinguish the effects of movement. It would
be like the theory of relativity without anything to relate to.

When Jason and the Argonauts are about to set out on their voyage to seek the Golden Fleece, Orpheus sings a creation story that
tells how "at first the earth, sea and sky are confused together in one
indistinct form."[8] Orpheus describes this confusion as the *atekmartoi*
ocean (Greek, "without *tekmar*"—point of definition). He does so to
provide an image of formlessness upon which creation can be imposed,
or from which it can arise. What these myths are trying to tell us is
that the "unbegun" universe lacks not *content* but rather *discrimination.*
When Orpheus tells us the earth, sea, and sky are confused into one,
he is alluding to the fact that the primordial ocean already contains the

material necessary to produce these separate elements, but without any discrete identity for any of them.

The same idea is described in a variety of ways in different traditions, but always with the same underlying implications. In the Babylonian creation epic, Enuma Elish ("When on High"), the primeval ocean is described as consisting of two halves, Tiamat ("the Depths") and Apsu ("the Abyss"). Although these two halves were conjoined at the beginning into an undifferentiated whole, we are told that they are destined to give birth to all that is to come:

> *When on high the heaven had not been named*
> *Firm ground below had not been called by name,*
> *Nought but primordial Apsu, their begetter,*
> *and Mummu-Tiamat she who bore them all*
> *Their waters co-mingling as a single body.*[9]

The image of the primordial waters has now gained a further significance. What was a vast and pointless waste has suddenly revealed itself to be pregnant with possibility. Mircea Eliade put it briefly: "Water symbolizes the whole of potentiality, the source of all possible existence."[10]

What this means is that whether preexistence is referred to as a formless ocean or as "the Void," or, as the Norse myths call it, the "Yawning Gap," it is by no means empty, although the suggestion that it contains nothing may lead some commentators to describe it as such. The nothing that it contains is to be regarded as no-thing, that is, no one thing differentiated from another. It is this absence of definition or delineation that makes it "indiscernible, unknowable."

The paradoxical relationship between nothing and no-thing is what the Rig Veda is describing when it talks of there being "neither non-existence nor existence." The raw material of preexistence is not boundless nothing but instead boundaryless everything. It is as if the myths are trying to talk about a world in which consciousness has not yet got to work. There has as yet been no identifying of the boundaries between this and that.

The process of division by means of establishing boundaries is one that the Jungian psychologists regard as fundamental to the act of "individuation," to the emergence of a fully realized consciousness. Carl Jung argued that "the unconscious shares its undividedness with the divinity (*'les extrêmes se touches'*) but as it manifests itself the opposites are split asunder, as the discriminating nature of consciousness requires."[11]

Indeed, it might be argued that the process of identifying boundaries is the essence of consciousness, for without such a primal act, organization and order are inconceivable. It is this process that plays the determining role in the story of creation, as Jung himself suggested. The Greek poet Hesiod, who was the first to use the word *chaos* to describe the ocean of potentiality, also gives us a picture of the act of creation. Chaos, he said, was a "confused tangle of roots, the springs, the extremities of all things, which, by becoming distinct from each other, are to produce the organized cosmos."[12]

THE MATTER OF TIME

Hesiod's description of chaos provides us with the clue to the whole process of creation. It is to be understood as the emergence not only of the distinctions between one thing and another, but of the organization of these distinguished things into a coherent picture as well. We might almost say that the whole process of creation is one of making sense out of confusion. This is what produces the cosmos. In fact, the Greek word *kosmos* implies "order," the correct arrangement of things in relation to each other.

It is the absence of this organization that characterizes Hesiod's chaos, and that is emphasized by the Latin poet Ovid in the lines that head this chapter. However unpromising Ovid's "undigested lump" may make it seem, this formless preexistence, with the "jarry seeds" it contains, will become nothing less than the body of the supreme god, the source of all becomings.

In third-dynasty Egypt, the ocean of chaos is said to have contained "a conscious principle in existence since antiquity began, the god Atum," and this god's name is said to mean "wholeness, completeness" ("which stresses," says Wallis Budge, "his metaphysical nature"). Somehow, in an act that the myths describe as "awakening," the formlessness of chaos is transformed, first into a sea of potential and then into the "supreme being," from whose complete and eternal body creation will emerge:

> In the beginning the universe was shrouded in darkness, total, indiscernible, undiscoverable, unknowable, as if it was completely absorbed in sleep. Then the Lord revealed himself, irresistible, self-existent, subtle, eternal, the essence of all beings.[13]

There seems to be no surviving mythology that does not know of this idea of the one supreme being from whom all things flow (although in some traditions it remains unknown to all but the highest initiates). In every case, this god has the same character. It exists eternally as a completeness, embracing in a timeless state all that can ever come into being, undivided and perfect. In many cases, this completeness has exchanged formlessness for the perfect form and is frequently pictured as spherical in shape and infinite in extent. As such it appears, whether it is the two calabashes, joined at the rim, kept in Yoruba temples that represent the deities Obatala and Odudua, Heaven and Earth, before their separation,* or the "divine All," the spherical cosmic god mentioned by Xenophanes of Colophon, a Greek poet writing around 500 BCE, who described it as being "similar on all sides."†

The perfect and undivided nature of this sphere was stressed by

---

*As described by A. B. Ellis in *Yoruba-speaking Peoples of the Slave Coast of West Africa*, 1894; www.sacred-texts.com/afr/yor/index.htm.

†"The pantheism of Xenophanes saw the divine All as a sphere, and the aura of divinity still clings to the spherical shape in Aristotle" (Guthrie, *History of Greek Philosophy*, 114); cf. Marie-Louise von Franz, *Number and Time*, 177: "On the whole, the most ancient thought patterns concerning the totality of existence were mandalas . . . frequently models of an infinite sphere. An ancient hymn extols Zeus as the first and last, 'the depths of the earth and the star-sown heaven . . . the great body of the king, in which all that exists rotates.'" Likewise, Parmenides described this deity as "everyway like unto the fullness of a well-rounded sphere evenly balanced from the centre on every side."

Anaximander (ca. 560 BCE), who used the Greek term *apeiron* to describe it, implying the "boundless," "to indicate that no line could be drawn between part and part within the whole." A Hindu hymn to this supreme spherical deity proclaims, "You stand changeless, unsullied; the supreme stage is your core, the universe is your shape."[14]

More-elemental traditions describe this perfect form less subtly, but more expressively, perhaps, of the fullness of potential, as an egg. According to Chinese tradition, "In former times, before heaven and earth existed, chaos looked like a hen's egg,"[15] and this egg will split apart and reveal its contents to be heaven and earth. Other traditions make this egg the home of the original god. In the Society Islands of Polynesia we hear how Ta'aroa, ancestor of all the gods and creator of the universe, "sat in his shell in darkness from eternity. The shell was like an egg revolving in endless space."[16]

Egyptian coffin texts from around 2250 BCE sum up the various aspects of this image, quoting the god Atum, the "Completeness":

> *I was the Primeval Waters*
> *he who had come into existence as a circle*
> *he who was the dweller in his egg.*
> *I was the one who began (creation)*
> *the dweller in Primeval Waters.*[17]

And the principal characteristic of these waters, according to Mircea Eliade, "is to precede creation. . . . [Water] can never get beyond its own mode of existence—can never express itself in forms. . . . Water can never pass beyond the conditions of the potential of seeds and hidden powers. Everything that has form is manifested above the waters, is separate from them. On the other hand, as soon as it has separated itself from water, every 'form' loses its potentiality, falls under the law of time."[18]

We can now return to St. Augustine's response to the question of what existed before creation, that there was no before, since God created time alongside the universe. In this he was echoing Plato's description of how "before the world's birth . . . there was no Time, for neither was there arrangement, measure or mark of division, only an indefinite motion, as it were the unformed, unwrought matter of Time."[19]

The mystery of the awakening, the appearance of the eternal body of God from the formlessness of chaos, is one of those elements in the myth that cannot be understood by the application of our usual cause-and-effect linear logic. As will become clear, this awakening is the result of light shone back by what is to follow. Indeed, it is hard to imagine "completeness" without at least some idea of what the separate elements are. These elements, whose roots lie in a confused tangle in Hesiod's chaos, are the potential that is said to be released by the appearance of the supreme god. They represent those things that are to emerge in accordance with the passage of time. Whether a tradition speaks of creation as arising from the formlessness of the waters or from the boundless spherical body of the one eternal God, it is the appearance of the boundaries from which time can be measured that is crucial.

When the divine eternal energy begins to operate, it is organization according to time that appears. The "things" whose "jarring seeds" abide ill-joined in Ovid's "barren load" are the elements that go in to making up the process of time. The boundaries that need to be distinguished for the world to emerge are those that allow the measurement of time. Any other organization of the world can exist only within a delineated framework for the operation of the machinery of time, and it is with the construction of this framework that the myth of creation is concerned.

The revolutionary achievement of Santillana and von Dechend in their book *Hamlet's Mill* was to show how this framework can be recognized in the various encrypted forms in which it appears in myth. Before they had offered their insight, such revealing statements as Plato's regarding the "unwrought matter of Time," despite their clarity, had had no context in which their meaning could be fitted. As a result, that meaning was inevitably misconstrued.

What had been lacking was the possibility that an understanding of astronomy sufficiently profound could have been available at such an early date. At such times, so the common assumption went, all aspects of human development were as little "advanced" as technology was. Not even the descriptions of the complex cosmologies and starry concerns of those untouched tribal societies so much discussed by the anthropologists could dislodge the conviction that the Stone Age mentality was little more than a bewildered stumbling toward intelligence. A further problem seems to have been that many commentators on the mythical material lacked sufficient knowledge of astronomy themselves to enable them to recognize it even if they had seen it.

It was the very unintelligibility of the material that first led Hertha von Dechend to her extraordinary revelation. In the end, despite her resistance, it had proved unavoidable. Once the leap of imagination had been made, once it was allowed that sophisticated astronomical knowledge might have been achievable, incomprehensible and apparently fantastic stories began to fit into the picture with astonishing lucidity. When the astronomical template is laid over even the most fanciful material, the formality of the patterns that are displayed is stunning.

It is the consistency and the wide-ranging nature of these patterns that provide the most convincing evidence that, despite the revolution it causes for our notions about human prehistory, such knowledge was developed. And if this is so, if the pursuit of the processes of time really did become so compulsive, there must have been a reason why.

It is clear, then, that any attempt to understand the meaning of myth and the function it serves must begin with an understanding of the measurement of time that it encapsulates. It should also be apparent that this is not just the measuring of the hours; the ancients were not concerned with finding windows in busy schedules. What dominated their intellectual lives was a preoccupation with cosmic time on a grand scale.

A COMPASS
UPON THE DEEP

She standeth in the top of the high places,
by the way in the places of the paths
The Lord by wisdom hath founded the earth.
By understanding hath he established the heavens
by his knowledge the depths are broken up
When he prepared the heavens
I was there when he set a compass upon the face of the depths.
PROVERBS 8:27

STRETCHING THE LINE

If the uncreated universe is viewed as a pathless ocean or a trackless waste, then the act of creation, the establishment of definition within the undefined, must involve the making of a pathway across it. In the song of creation that Orpheus sings to send the Argonauts on their way,[1] he describes how the routes across the boundless ocean are first established:

At first the earth, sea and sky are confused together in indistinct form. . . . Then as they emerge from the primordial chaos the stars, the routes of the moon and sun, form in the sky the *tekmar* which is thenceforward fixed. . . . The movements of the stars, sun and moon, by which the whole course of Becoming is regulated . . . are the vis-

24

ible routes [Greek, *hodoi*] which plot out the various regional spaces. By means of their paths, thus harmonized, they establish the East and the West, the North and the South, thus giving orientation to a space which without them, would remain formless and undefined.[2]

What Orpheus describes as the emergence of the stars from the primordial chaos is the formation of an understanding of the patterns of movement visible in the vault of heaven. These patterns, the regular periods of the stars and the passage of the planets through them, laid out the routes across the waste.

It was the order visible in the night sky that generated the mythic cosmology according to which the whole of creation proceeded. So profound was the effect of the order revealed by the regularity of these patterns, so awe-inspiring in their apparent completeness and precision, so immune to the machinations of human observers, that they were read as being a broadcast of the "laws" in accordance with which the universe functioned.

In this way were made those marks of division on the unwrought matter of time that Plato had said were missing "before the world's birth." Here began the creation of the cosmos. The myths of creation, the formative repository of this cosmos, thus became an expression not only of the order of the physical world, of the laws of nature, but also of the laws of all that had existence within it, including human society. All human activity would therefore ideally be ordered according to the patterns given form by the myths. In the introduction to *Hamlet's Mill*, Santillana discusses the ways in which the knowledge of astronomy and its role in mythmaking has long been lost. He explains that today's experts

tell us that Saturn and Jupiter are names of vague deities, subterranean or atmospheric, superimposed on the planets at a "late" period . . . all unaware that planetary periods, sidereal and synodic, were known and rehearsed in numerous ways by celebrations already traditional in archaic times . . . Ancient historians would have been aghast had they been told that obvious things were to become unnoticeable.

Aristotle was proud to state it as known that the gods were originally stars, even if popular fantasy had later obscured this truth. Little as he believed in progress, he felt this much had been secured for the future."[3]

The gods were originally stars. The statement is so incomprehensible to us, so alien to the way in which we have been taught to think about these matters, that we are unable to imagine what it might mean, let alone assess its possible truth. And yet Orpheus tells us that it is the movements of the stars that give form to chaos. They regulate "the whole course of Becoming," he says, repeating a doctrine well known to the Egyptians. A hymn to the Egyptian Ra says:

> *Thou art creator of the starry gods in heaven above*[4]*
> *Homage to thee O lord of starry deities.*[5]

As we begin to explore the manner in which this regulation is achieved, Aristotle's statement will begin to make more and more sense. The determination of the laws of creation is, after all, a very suitable role for the gods.

How, then, do the stars play this role? How do they impose the marks of division on Plato's unwrought matter of time? For the beginnings of an answer to these questions, we can turn to an illustration from the mid-seventeenth century that Jung reproduced in his work *Psychology and Alchemy*.[6] It is entitled "The Unfettered Opposites in Chaos." It shows the zodiac constellations, grouped in pairs of opposites, swirling in a dark mass of cloud, wind, fire, and water.

---

*Throughout this book, the translation of the *Egyptian Book of the Dead* that I refer to is that by E. A. Wallis Budge (London: Arkana, 1989). The British Museum says of Wallis Budge's books in general, "They are at best unreliable, and usually misleading." I have therefore compared his translations with those of R. O. Faulkner (*The Ancient Egyptian Book of the Dead,* London: British Museum Publications, 1985) and made reference to differences where they seem relevant. My approach has been based on the critical adage *difficilior lectio potior est,* "the difficult reading has power."

It is worth recalling Jung's description of the undividedness of both the deity and the unconscious. The process of the manifestation of consciousness—which, in Jung's opinion, mirrored the creation—involved what he called the "separation of the opposites." In this illustration we can see just how these "opposites" are to be understood. Hidden among the swirling shadows and darkness can be discerned, combined in pairs but without order, the symbols of the zodiac.

Whatever the actual age of the familiar symbols of the zodiac,* it is clear that the concept of a path across the sky whereon the planets travel is an ancient one, and also the notion that the stars that lie along this path, in positions that are fixed relative to one another, can act as markers to divide it up into sections. Principal of the "planets" that wander this starry path is, of course, the sun. In the Vedic tradition of India, it is sometimes the god Varuna, sometimes Vishnu, who measures creation: "This great feat of the famous Asurian Varuna I shall proclaim who stood up in the middle realm of space and measured apart the earth with the sun as with a measuring stick."†

This measuring line must be related to that cord with which, according to the Rig Veda, the poets measured off existence from nonexistence:

> *There was neither existence nor non-existence,*
> *there was neither the realm of space nor the sky which is*
> *beyond. . . .*
> *Poets seeking in their heart with wisdom*
> *found the bond of existence in non-existence. . . .*
> *Their cord was extended across.*[7]

*Throughout this work, the term *zodiac* is used as shorthand for the various collections of stars that lie along the path of the sun. This is not to imply everywhere and always the same groupings and names as we now have, but simply to represent the concept of a path defined by these stars. Nor should it be taken to suggest a date for the first naming of those signs we now recognize.

†Wendy O'Flaherty, *Rig Veda*, 211, Hymn 5, 85.5. cf. 226 RV.1, 154.1: "Let me now sing the heroic deeds of Vishnu, who has measured apart the realms of the earth, who propped up the upper dwelling place." O'Flaherty points out that the verb *prop* is *skambh*, "related to the noun for *pillar*."

The word used here for *poet* is the same as that for *measure* (in fact, the word is *kavi;* see the section titled "Forging the Sky," page 60). In another hymn (Rig Veda 1.160.1), the sun itself is referred to as "the poet of space." The act of stretching the measuring line is fundamental to the process of imposing order on chaos. Thus, the God of the Hebrew scriptures challenges Job to explain the work of creation by asking, "Who decided the dimensions of it, or who stretched the measuring line across it?"[8]

What these stories are describing is the use of the regular passage of the sun across the stars as a means of measuring time. As is well known, this annual journey takes place along the same path every year, and the stars that mark this path are those that today make up the figures of the zodiac. The first mark of definition imposed on the matter of time is the point on this journey that the sun occupies at the beginning of the year. Most traditions recognize the division of the path of the sun into the familiar twelve 30-degree "signs," although the precise locations and the nature of the figures used to describe them vary across the world, and we frequently find a division into twenty-seven or twenty-eight "signs" based on the path of the moon. The establishment of the role of these stars as markers is what enables the cycles of time to be measured. Without the knowledge of their pattern, the sky would indeed be a vast and pathless ocean—chaos, in fact. Once the positions of these markers relative to each other have been established, the stage is set for time to begin. The complete circle of the sun's path, referred to by astronomers as the ecliptic, becomes a form of measuring rod on which the course of time will be plotted.

THE SERPENT CIRCLE

In the early traditions of Persia, there appears a deity of infinite time named Zurvan Arkana. In several depictions he is represented as coiled about by a serpent and decorated with the constellations. The Jungian psychologist Marie-Louise von Franz remarks, "Here the circle of the zodiac is identical with the Worm Ouroboros, the serpent in matter, a symbol of the all-encompassing unity of the cosmos."*

Not only the unity of the cosmos but also its very existence depend on the embrace of this serpent. In West Africa, the Fon people tell of the rainbow-serpent Dan Ayiedo Hwedo, tail in mouth, coiled beneath the earth to stop it from sinking. His coils, revolving above the earth, keep the planets and the stars in motion. If they were loosened, the world would fall apart.

This is the sort of story that, extracted from its context and presented as "primitive belief," contributed to the idea that stories told by myth dated from the childhood of mind. Nothing could be further from the truth. The story appears around the world, as if this globe-embracing, world-supporting serpent could be seen clearly by anyone anywhere who cared to look. It is pictured on a number of runic stones from Sweden, dated around 1000–1100 CE. The one on Froson Island tells the story of a serpent so long that it reached around the whole island and was tied head to tail by a ribbon. Were the ribbon to break, the island would sink.†

In Norse tradition, the encircling serpent is called the Terrible Girdler of the Earth, the Midgard Worm. In India, the smith Tvastra, the "fashioner of the gods," built himself a house and created Vritra, the serpent monster, as a sort of roof, but also as walls for his habitation: "Inside this dwelling, encircled by Vritra, Sky, Earth and Waters existed."[9]

Supporting the creation is not the only role for the serpent, however; many traditions hold that it had been present during the act of

---

*Marie-Louise von Franz, *Number and Time* (London: Rider, 1974), 179. Von Franz is here referring to the image of the *ouroboros* included in the picture of the "so-called Chrysopoeia of Cleopatra" of the Greek alchemists. Many images exist that link this serpent with the signs of the zodiac; see, for example, C. J. Jung, *Psychology and Alchemy*, 191. In the Gnostic text Pistis Sophia, this is the dragon of outer darkness "whose tail is in his mouth, outside the whole world and surrounding the whole world." This dragon has twelve "dungeons" in which the souls of the damned exist until "the ascension of the universe."
†A picture of this stone, and the accompanying story, appears in J. B. Sweeney, *A Pictorial History of Sea Monsters* (New York: Nelson-Crown, 1972), 12.

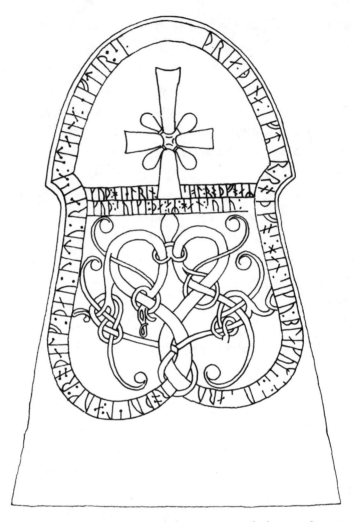

Fig. 1. The Binding Serpent. Detail from an inscribed stone from Sjonhem Parish Church, Gotland, Sweden.

creation itself. The Greek poet Hesiod tells how the serpent Okeanus (or Ophion), who resides at the very edge of the universe, was at hand when Thetys, the goddess of the depths, first created heaven and earth. The Popol Vuh, the creation text of the Maya, confirms the primeval nature of this cosmic serpent, telling us how, at that time when "the face of the earth was not to be seen, only the peaceful sea and the expanse of heaven, there were only immobility and silence in the darkness and in the night alone was the Creator Tepen and Gucumatz the Plumed Serpent."[10]

The Australians also knew of this primordial serpent. In the north-western region, the Unambal people called it Ungud: "In the beginning only earth and heaven existed. Deep in the earth lived Ungud, in the form of a Great Snake (often identified with the earth but also with the waters). In the sky lives Wallanganda, Lord of heaven, but at the same time, The Milky Way."[11]

In India the serpent is Sesa, whose coils symbolize "the endless revolutions of Time." Sesa's abode is in Nara, the primal waters where also resides the god Vishnu, in the form of Narayana, who is "this primeval universe. He at the time of creation makes the creation and at the time of world-destruction he devours it again."[12]

Between times, Narayana sleeps upon the coils of Sesa. These three, Narayana (Vishnu), the serpent Sesa, and the waters, "are the triune manifestation of the single divine imperishable cosmic substance, the energy underlying and inhabiting all forms of life."[13]

Probably the earliest recorded version of the story of the World Serpent appears in texts from Egypt. There, the god Atum, who had been the primeval waters before he came into existence as a circle, also identifies himself with the "great serpent of the Deep," describing himself as "first and last Image of the god Atum. . . . When earth will become again part of the Primeval ocean . . . then I will be what will remain.* . . . I will have changed myself back into the Old Serpent."[14]

The implication of all these stories is that the circle of the stars that make up the zodiac, from which the organized earth is measured off, in some way represents the body of God. Not only are the stars to be recognized as gods (or as the Hebrew scriptures call them, the sons of God, the heavenly host), but in some way they are also to be taken all together as the body of the eternal God. Exactly how this notion of the eternal and undivided circular body of God acts to create the world

---

*The Hindu serpent Sesa's name means "remainder." It is also known as Ananta ("endless").

Fig. 2. The Coils of Time. Vishnu asleep on the coils of the serpent Sesa. Detail from Atha Naradiyamahapuranam (Narada's Great Ancient Tale), Mumbayyam: Sri Venkatesvara Stim-Yantragare, 1923.

will become clearer. That it was once identified with the wheel of the zodiac was still hinted at by the alchemists when they called the circle described by the sun the "line that runs back on itself like the snake that with its head bites its own tail, wherein God may be discerned."[15]

THE WHEEL OF THE GODS

In the Zend Avesta scriptures, the Persian deity Zurvan, when he appears encircled by the serpent, is identified with the "the invisible intangible primal matter, uncompounded and devoid of parts,"[16] that is to say, he represents the eternal completeness, the source of being. But he is also called *ras*, "the wheel," the word used elsewhere to describe the wheel of heaven. In Hindu lore, this wheel is the "kalacakra—the wagon-wheel of Time, . . . in whose revolution all creation has its being and sees its end."[17]

The imagery of the wheel is developed in the Rig Veda hymns into that of the "one-wheeled chariot," an image of the whole of creation. The hymn known as the "Riddle of the Sacrificer" describes how "all these creatures [all the elements of creation] rest on the ageless and unstoppable wheel. . . . The twelve-spoked wheel of Rita ["order"; see "The Queen of Heaven," page 145] rolls around and around the sky and never ages. . . . All the worlds rest on this five-spoked wheel that rolls around and around."[18] According to the *Dictionary of Hinduism*, the word used here is *cakra* ("wagon-wheel"), which can be a "mystical circle or diagram" but is also "the stellar *cakra*" composed of the *nakshatras* (the twenty-eight star groups that divide up the journey of the moon as it travels along the ecliptic) and the planets. This cakra is said to be "regarded as rotating like a potter's wheel." The *dharma-cakra* is the "eight-spoked wheel of the law," said to represent celestial balance and cosmic order. Brahma, too, is described as holding the universe together in the manner of the axle, spokes, and felly of a wheel.

These ideas of the whole of creation likened to a rotating wheel survived into the era of the medieval alchemists, as Jung displays in his work on psychology and alchemy. Having mentioned the connection between the wheel of creation and that of the zodiac, he describes its relationship to the work of the alchemists: "This wheel has a significant connection with the *rota* or *opus circulatorium* [the circular work] of alchemy. As Dorn says [in the "Philosophia Chemica"]: 'The wheel of creation takes its rise from the *prima materia,* whence it passes to the simple elements.' Enlarging on the idea of the *rota philosophica* [the philosopher's wheel], Ripley says that the wheel must be turned by the four seasons."[19]

The wheel thus turns into the wheel of the sun rolling around the heavens. In India too, the most common interpretation of the cakra was as a "symbol of the sun," and it is with the so-called sun-wheel that we are most familiar, although this image has been much misunderstood. It should not be regarded as representing the sun as such, however much that body may be said to "roll around heaven all day." It is, in fact, an image of the passage of the sun through the stars, its spokes marking out the succession of "mansions" into which that passage is divided.

We should be careful not to jump to the conclusion that this wheel also represents the year, however—at least, not in the literal sense. That something more profound is involved is suggested by the Hindu text, which refers to the year as "the revolving wheel of the gods which contains everything including immortality. . . . On it the gods move through all the worlds."[20]

In fact, the Hindu priests knew of three "mystic wheels of the sun." The sun's yearly path through the zodiac is the second of these, the first being the sun's daily path across the sky. The third wheel, however, is invisible and is revealed only to those "who are skilled in the highest truths." The Rig Veda hymn to Surya, "the sun," says, "Your two wheels, Surya, the Brahmins know in their measured rounds. But the one wheel that is hidden, only the inspired know that."[21]

To those who studied the night sky to learn the laws of the universe was revealed the greatest of all secrets. They slowly learned that the circle that they were studying and upon which they were modeling existence was only a little wheel within a much greater wheel, a wheel that was great enough to deserve the name of the Eternal Body of the Supreme God.

Now, it may be that in some of the myths that we know of are recorded models from before this moment of discovery, but it seems to be this revelation above all others that has given rise to the mythmakers' boldest attempts to order experience, and that gave the myths of creation the determinative form in which we have received them. That such a revelation was discerned at all is evidence of the diligence and expertise with which these observers carried out their work, for only in this way could they have become aware of the fact that this model, so majestic in its precision, contained one dreadful secret. So grindingly slowly does such a secret display itself that any one observer might witness only the slightest evidence of it in a lifetime. And yet it was a secret whose awful dawning would be cataclysmic for the model they were constructing and upon which they had based the structure of their lives, a secret that was to transform their notions of self and cosmos in a way that has reverberated ever since.

Out of this secret and its consequences have been constructed not only the notions of eternity and immortality but also the language and imagery for the whole body of philosophy concerning being and becoming, the relationship of humankind with the Godhead, and the beginning and end of the world. And the consequences of this secret still echo unacknowledged through all the vast architecture of our language, our institutions, and our most enduring aspirations.

To restabilize their model of the world after their cataclysmic discoveries about the processes of the heavens, the early mythographers had to include the invisible third wheel of the sun. To describe its awesome significance for their cosmos, they began to tell the story of how the body of God was separated from eternity and divided into the forms of creation concealed within this invisible wheel. It is nothing less than the eternal wheel within which all the lesser wheels turn and from which they take their motion (as Ezekiel saw it in his vision). To learn the secret of this wheel, it is indeed necessary to become "skilled in the highest truths."

HOW THE EARTH
WAS MADE

hoy wakah chanal waxakna-tzuk u ch'ul k'aba yotot xaman
[it was made proper, the raised up sky place,
the eight-house partition is its holy name,
the house of the north]

MAYAN INSCRIPTION DESCRIBING THE
RAISING OF THE WORLD TREE,
OR *WAKAH CHAN,* "THE RAISED-UP-SKY"

THE FIRST EARTH

Within the waters of the primeval ocean lies coiled the serpent who is
the completeness of time. Waters, Vishnu, and Ananta (another name
of the serpent, which means "endless" or "eternal"), these three together
represent the aspects of the unmanifested Absolute. But this completeness cannot become the manifest world, as Eliade observed: "Water can
never express itself in forms. . . . Everything that has form is manifested
above the waters, is separate from them."[1]

The separating out of form from formlessness is the essential procedure of creation. The Hindu traditions hold that the first manifestation
came about when from the navel of the sleeping Vishnu a pink lotus
flower emerged. Seated at its center is Brahma, the primeval ancestor
with four arms, each holding one of the sacred texts of the Vedas, but

not even he can discover why this lotus flower came about, and what it is fixed to.

Across the world less erudite but more informative versions of this first rising from the waters occur. The Samoyed Yuraks, of the far north of Europe, for instance, tell how the supreme god, Num, seeing only water all around him, sent several birds in succession to explore the watery depths, until one of them, a diver bird, returned with a small fragment from the bottom in its beak. From it Num consolidated a floating island, and so was established the First Earth.

In case this story appears as one more "primitive" tale, Mircea Eliade supplies a more generalized version: "A cosmogonic myth brings on the stage the primordial waters and the Creator, the latter either as anthropomorphic or in the form of an aquatic animal, descending to the bottom of the ocean to bring back the material necessary for the creation of the world. The immense dissemination of this cosmogony and its archaic structures point to a tradition inherited from earliest prehistory."[2]

As we have already seen, traditions inherited from "earliest prehistory" are seldom as "primitive" as they may look. As for this island of mud, upon it the whole of organized creation is to be founded. A story from the tribal cultures of central India actually combines the story of the lotus and the diving animals:

> In the beginning nothing existed but the waters and a lotus with its head above the waters. Singbonga, the supreme spirit, lived in the lower regions. He came up to the surface by way of the hollow stem of the lotus and sat down on the lotus flower. Then Singbonga decided to create the Earth. He ordered the tortoise to seek him out some mud at the bottom of the ocean. But the tortoise failed in this undertaking and then the crab failed likewise. Finally the leech dived down, swallowed some mud, surfaced and disgorged the mud into Singbonga's hand. Then, on the surface of the water, Singbonga formed a continent bordered on its four sides by four seas. Then he sowed seeds, and trees grew.[3]

The fact that in all these stories there is in existence a whole collection of creatures before the establishment of the First Earth should lead us to suspect that something other than the creation of the world as we normally understand it is being described. Indeed, some versions of the story, such as those from North America, describe this kind of creation as occurring after the flood. (In this case, the persistence of the diving animals, particularly when they are birds, inevitably calls to mind Noah's doves.)

The Greeks used the word *apoiron* to describe a ring without a stone or setting, meaning "without path or division," a word also used to describe the uncharted seas and chaos itself. The appearance of the "solid earth" represents the first *tekmar*, or point of definition, in the abyssal sea, a stone set in the ring of immeasurability. From it, measurement can commence, and it will become the center of the structure of the universe. This foundation stone later found its way into the mysteries of the alchemists as the "philosopher's stone"; it had powers of regeneration as a result of its original creative role, and it is referred to in Revelation 2:17, where John is told, "I will give him a white stone with a new name written on the stone which no-one knows except him who receives it."*

According to Japanese tradition as recorded in the *Nihongi* (one of the *Chronicles of Japan,* written down in 720 CE), this First Earth has a slightly different origin, and from its story we learn of the next stage in the tale of the emergence of order from chaos:

> In the beginning was chaos, like an ocean of oil or an egg, shapeless but seed-bearing. The Celestial Deities Izanagi and Izanami leaned over the floating bridge of Heaven and plunged the Celestial spear

---

*Cf. C. G. Jung, *Psychology and Alchemy,* para. 244, p. 178: "The stone is that thing midway between perfect and imperfect bodies." The stone appears again in the origin myths of several of the tribes of southeastern Borneo, as recounted by Roland Dixon in his *Oceanic Mythology* (1916). One version states, "In the beginning there were only the sky and sea, in which swam a great serpent upon whose head was a crown of gold set with a shining stone. From the sky-world the deity threw earth upon the serpent's head, thus building an island in the midst of the sea; and this island became the world."

into the ocean of Chaos. They stirred it till the liquid coagulated whereupon they withdrew the spear and the drops of brine which fell from it formed the island Ono-goro-jima. The celestial couple then went down to the newly-formed island and made it the "Central Pillar of the earth."[4]

Fig. 3. Izanami and Izanagi Stirring the Ocean of Chaos

PROPPING UP THE SKY

The first island thus becomes the foundation on which a central "pillar" is established. The principal role of this pillar is to hold up the sky above the earth, as we hear in the continuation of the Chinese story that described chaos as looking like a hen's egg. It goes on to tell of P'an Ku, who was born in this egg: "After 18000 years chaos 'opened up'; the coarse heavy elements, yin, formed the earth; by contrast the light pure elements, yang, formed the sky. Every day the sky rose by ten feet. Every day the earth grew deeper by ten feet. And every day P'an Ku found that his size had increased by ten feet, so that after 18000 years P'an Ku's body was as great as the distance between heaven and earth."[5]

P'an Ku's name is made up of two words, one meaning "bowl" and representing the upturned hemisphere of the sky, and the other "to make fast." Having opened up chaos, his role is to maintain the proper relationship between the earth and sky that results. He has his physical counterpart in those temple towers set up in Sumer and Babylonia and referred to as "lofty column[s] stretching up to heaven and down to the underworld, the vertical bond of the world."[6]

This pillar is also the famous "World Tree," the axis mundi, around which the world rotates. According to David Freidel, "The Classic [Mayan] texts at Palenque tell us that the central axis of the cosmos was called the 'raised-up-sky' because First Father had raised it at the beginning of creation in order to separate the sky from the earth."[7]

This "raised-up-sky" is the *Wakah Chan,* represented as a tree in the shape of a cross. The same concept, although without the cross form, was described by the Sioux shaman Black Elk, who called it *Waga Chun,* the cotton tree that he was told to plant at the center of "the hoop of the nation, where the two roads cross."[8]

In the Caroline Islands of Micronesia, the oldest god, Solai, plants a tree (or mast) on the primordial rock (considered to be the "matrix" of the universe, existing in an endless stretch of original sea). He then proceeds to climb up this tree, stopping halfway to create the flat plane of the earth, then continuing to the top, where he places the sky.[9] (Note

Fig. 4. The Raised-up-Sky. The cross at the Temple of the Cross, Palenque. The Mayan Wakah Chan, erected by First Father to separate sky from earth. Drawing by Linda Schele, © David Schele, courtesy Foundation for the Advancement of Mesoamerican Studies, Inc. (www.famsi.org).

that this primordial rock, which elsewhere is said to be the First Earth, has now taken up a position beneath the "flat plane of the earth.")

Many traditions have regarded this pillar as a mountain, whether it is considered to stand on the original island at the center of the ocean or more generally at the "navel of the Earth." Typical of these mountains is the Hindu Mount Meru, said to pass through the middle of the earth and protrude on the other side. On this mountain the Hindu gods reside, just as the Persian Ahura Mazda had his seat on Mount

Hara. The equivalent in the Hebrew scriptures is described in Psalm 48: "Great is the Lord . . . in the City of God. His Holy Mountain, beautiful in elevation . . . is the joy of all the earth. Mount Zion in the far north, the city of the Great King."[10]

According to David Freidel, the Maya called this mountain Yax-Hal-Witz, First True Mountain, which rises "directly out of the waters of Creation." A pyramid building discovered at Waxaktun and dating from the earliest Mayan kingships represents this world mountain. It contains a mask sculpture on its surface that personifies the mountain as a monster. In the mouth of this monster is a smaller figure with a mirror, the Mayan glyph for the word *partition*. "This word," says Freidel, "refers to the idea of the marking of the four corners or directions in relation to the centre."[11]

At the top of any of these various world mountains is to be found the polestar, regarded as the pin that holds the sky in place and about which it turns. It is often referred to as the North Nail.* Babylonian sculptures survive of figures holding huge nails, and these are described as "foundation gods," and so they are, since it is this nail from which the whole of creation depends, and on whose permanence the fate of the world hangs, as the prophet Isaiah recognized when he foresaw its downfall: "In that day . . . the peg driven into a firm place will give way. It will be torn out and will fall. And the whole load hanging on it will be shattered."†

Among the Samoyed Yuraks and other Uralian people, it was considered that the column that supports the sky had to be kept in a good

---

*Mircea Eliade, *The Myth of the Eternal Return,* 15 ff. Eliade gives a number of examples of the central cosmic mountain, including one from "Semang Pygmies" of the Malay peninsula. He also recounts (page 19) the description of how, before the foundations of a building are laid, "the astrologer shows what spot in the foundation is exactly above the head of the snake that supports the world. The mason fashions a little wooden peg . . . and drives [it] into the ground in such a way as to peg the head of the snake securely down. . . . If the snake should ever shake its head really violently, it would shake the world to pieces."

†Isaiah 22:25, Jerusalem Bible translation. The passage is traditionally held to describe the fall of the House of Eliakim. A "House" in this sense can be seen as a model of an "Age of the World." The church fathers saw in Eliakim a foreshadowing of the Messiah.

state of repair; otherwise the universe might collapse and the firmament fall and crush the earth. The polestar was presumed to be the top of this sacred column, and about it the sky turned. In the Hindu scriptures, the polestar is referred to as "the seat of Vishnu," and it is to this elevated position that Prince Dhruva, who "stood on one leg for more than a month," is said to have been promoted. As a reward for his remarkable feat, the gods announced that he should ascend "to the exalted seat of Vishnu, round which the starry spheres forever wander, like the upright axle of the corn-mill circled without end by the labouring oxen. . . . The stars and their figures and also the planets shall turn around you."*

THE BONDS OF HEAVEN

Our little mud island in the middle of the primordial sea has become something very much grander. Prince Dhruva acts as a kind of axle for the whole of heaven, an embodiment of the central pillar. At first sight, it looks as if the location of this central pillar, whether at its base at the "navel of the Earth" or at its summit at the "pivot of the sky," must be the first mark of division on the "unwrought Matter of Time." Mythical imagery is seldom so straightforward, however. To carry out his task of fastening earth and sky, Dhruva is said to exert pulls, referred to as winds, upon the stars, to keep them in their circlings. These "winds" have an essential part to play in the process of defining creation, but the Hindu text has little more to say that will clarify the matter.

For elucidation we can turn to the Egyptians, who considered the task of Prince Dhruva's winds to be accomplished by ropes: "The Great God lives, fixed in the middle of the sky upon his support. The guide ropes are adjusted for that great one, the Dweller in the Sky."[12]

The same idea reappears (or perhaps appears for the first time) in

---

*The text is from the Bhagavata Purana, quoted in Santillana and von Dechend, *Hamlet's Mill,* 138. For an image of Dhruva (Dhruga), see David Maclagan, *Creation Mythology* (London: Thames and Hudson, 1977), 49.

Australia, in a legend describing the cause of lunar eclipses. It tells how "Yhi, the sun, after many lovers, tried to ensnare Bahloo, the moon; but he would have none of her, and so she chased him across the sky, telling the spirits who stand round the sky holding it up, that if they let him escape past them to earth, she will throw down the spirit who sits in the sky holding the ends of the Kurrajong ropes which they guard at the other end, and if that spirit falls the earth will be hurled down into everlasting darkness."[13]

This concept is echoed among the Native Americans of California: "The earth is a great island floating in a sea of water, and suspended at each of the four cardinal points by a cord hanging down from the sky vault, which is of solid rock. When the world grows old and worn out, the people will die and the cords will break and let the earth sink down into the ocean, and all will be water again. The Indians are afraid of this."[14]

Once we understand that these ropes hold creation together, we can better explain why the Hindu deity Varuna, described in the Rig Veda as being the one through whom "the awful heaven and earth are made fast," is also described as the "Master of Bonds."

For a fuller picture of how these "guide ropes" or "bonds" hold earth and sky together, we can turn to Plato's *Republic,* in which he describes how Er and his companions, on their journey to the underworld, come upon a vision: "[They saw reaching] through the whole of heaven and earth as it were a pillar, for colour most like unto the rainbow. . . . And there at the middle part of the Light, [they] beheld extended from the heaven the ends of the bonds thereof; for this light is that which bindeth the heavens together; as the undergirths hold together ships so does it hold together the whole round of heaven, and from the ends extends the Spindle of Necessity, which causeth all the heavenly revolutions."[15]

The North Nail and the "girths" suspended from it are not sufficient to make fast the earth, however. To complete the structure, these girths need to be fastened to the "hull" of the sky. The fastenings themselves are described in the Rig Veda hymn addressed to the mysterious Skambha, "the world scaffold":

*Firmly did Skambha hold in place heaven and earth, both
     these.*
*Firmly did Skambha hold in place the six widespread
     directions,*
Whither go the half-months,
*whither go the months in ritual understanding with the
     year?*
Tell forth that Skambha which and what is he?
*On whom Prajapati propped up the worlds and sustained
     them all? . . .*
Two single maidens of differing form
Weave a web on six pegs set approaching it in turn
The one draws the threads the other sets them in their place
*. . . These pegs prop up the world.**

Of the "six widespread directions" held in place by these six pegs,
two appear to be up and down and two of the pegs themselves must
represent top and base of the central column. The other four pegs are
located at the "four corners of the earth," where Shu, the Egyptian god
of light, set up the four pillars to hold apart his parents, the earth and the
sky. The earth has become not only a flat plane but also a square one.

For the Maya, the supporting role of the pillars was played by the
babatunob, the sky-bearers. They appear at the corners of many temple
representations of the World Mountain, holding the heavens up above
the earth.[16] At the opposite extreme of architectural image, the Inuit
Indians depicted the earth as a "tent resting on pegs, with a cover over
it—the vault of heaven—which was slashed by a knife in four different places to allow the north, south, east and west winds to escape."
And in case we should mistake these "winds" for their meteorological

---

*R. C. Zaehner, *Hindu Scriptures,* Atharva Veda 10.7.35–44. Another translation of
this important text is at www.vedah.com/org/literature/atharvaVeda. Zaehner also gives
these verses from Atharva Veda 10.8: "By Skambha are these two held apart, heaven and
earth . . . Twelve fellies, a single wheel, three naves. Who can understand it? . . . Three
hundred pegs have there been hammered in and sixty nails which none can move."

equivalents, even after we have learned about Prince Dhruva's "winds," the prophet Enoch recorded a vision he had of them. With a logic typical of myth, they are also pillars: "I also beheld the four winds, which bear up the earth, and the firmament of heaven arising in the midst of heaven and of earth and constituting the pillars of heaven. I saw the winds which turn the sky and which cause the orb of the sun and of all the stars to set."[17]

The imagery of these pillars is as diverse as it is widespread. The Rig Veda refers to them in hymn 1.160: "He measured apart the two realms of space with his power of inspiration and fixed them in place with undecaying pillars."[18]

The serpent known by the Fon people of West Africa to be keeping the stars in motion by means of his coils is said to have built four pillars to support the heavens, twisting himself around them to keep them up. Monkeys feed him iron bars; if they were to fail in this task, the pillars would fall and the earth would collapse. References to the pillars that support the plane of the earth also appear in the Hebrew scriptures, although they are sometimes obscured by translation. The term translated in the King James Bible as "foundations," particularly in the Book of Job, is given as "pillars" in the Jerusalem Bible. In a footnote to Psalm 46, the Jerusalem Bible describes how "the earth rests on the waters of the nether ocean [Psalm 24.2], being supported by pillars [Psalm 75.3]."

Both the King James and the Jerusalem translations agree on the pillars referred to in 1 Samuel 2:8: "For the pillars of the earth are the Lord's and he hath set the world upon them."

THE MEASURES OF CREATION

Great is the recital, the history of the time
when all the corners of the sky and earth were completed,
and the quadrangulation, its measure, the four points,
the measuring of the corners, the measuring of the lines,
in heaven, in earth, at the four corners . . .

FROM THE MAYAN TEXT OF
THE POPOL VUH

SURVEYING THE WORLD

The Mayan celebration of the establishment of creation is a description of the process of surveying. A more recent translation of the Popol Vuh verse quoted above makes this even clearer:

The fourfold siding, fourfold cornering,
measuring, fourfold staking,
halving the cord, stretching the cord
in the sky, on the earth,
the four sides, the four corners as it is said.[1]

The pillars that support the sky are to be regarded as benchmarks, boundary stones put down to enable an order to be imposed upon chaos, and a measure to be made of the earth. Wherever a count is made of these pillars, there are always four. It should be apparent after all this discussion of the corners of the earth, however, that it is not the earth we walk about on that is of concern here. This particular "earth" is not only square but flat as well. Exactly what this means will soon become clear. We should recall that it is the matter of time that is being measured in this surveying process. This is the process that the prophet Enoch describes when, during his vision, he asks his guide to explain the actions of the angels they see: "Wherefore have they taken those long ropes and gone forth?" The guide responds, "They are gone forth to measure. These measures shall reveal all the secrets of the depths of the earth."[2]

The process these angels were going forth to conduct must have been similar to the ceremony that in Egypt was known as "stretching the rope" and that accompanied the commencement of any building project, thus relating its construction to the original founding of the earth. Its purpose is to establish the first point of reference from which the circle of the stars can be measured, to divide the sky.

In the Book of Job, in the Hebrew scriptures, God challenges Job to answer: "Where wast thou when I laid the foundations of the earth? . . . Who decided the dimensions of it . . . or who stretched the measuring line across it?"[3] This process of surveying the foundations of creation reappears in the verses of the Rig Veda that tell how Varuna made fast the earth and heaven, and here we are told specifically that he did so by using the sun as a measure, and as a commentary on these verses points out, "We must take this measure as a measuring line."

William Blake's depiction of the creator with his compass is perhaps the most familiar image of this notion of the creator as surveyor. As if to emphasize the importance of this role, the two Chinese deities responsible for creation come equipped with compass and plumb line.

What all this means is that the "bindings" that hold heaven and earth together are also to be taken as plumb lines that stretch out the

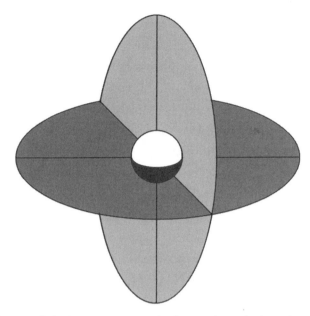

Fig. 5. The Divided Sky. A Great Circle drawn through the poles provides the first measuring line across the circle of the stars.

Fig. 6. The Royal Surveyors. Detail from a rubbing from a stone engraving in the tombs of the U family in the province of Shantung, China, second century CE. Fu-Hsi (right) and Nu-kua (left). Fu-Hsi was the first to exercise sovereignty; the inscription tells how he set out the trigrams and knotted ropes in such a way as to ensure good government of the lands enclosed by the sea.

measures of creation across the circle of the stars. It is a particular state of order that is being held fast.

Representations of the cosmos constructed in this way appear again and again, often in unlikely places. An elemental one is the rectilinear cross that appears on the English flag (the Royal Orb is a more complete image, including as it does the spherical universe bounded by the cross). Another, more detailed example from China comes in the form of a series of mirrors known as the TVL mirrors because of their decorative patterns, described as follows:

> The four T's that stand around the square earth are the four cardinal points. These lie at the extremities of the two cosmic lines or ropes which cross the universe and hold it together; the T's comprise a vertical prop which supports a horizontal beam; the four V's mark the four corners of heaven. The four L's represent devices used by carpenters to set a straight line . . . thus they symbolize the ends of the two cosmic lines as if they had been drawn with the help of such an instrument.[4]

For the Mayan people today, this kind of creation is still repeated as part of ceremonies to restore the correct balance of the world. David Freidel describes a ceremony he witnessed to bring to an end a prolonged drought. The officiant constructed an altar consisting of a table overarched by four saplings bent to intersect above its center. From this intersection was hung a circular platform referred to as the "the sky-platform." Sturdy vines of a kind called *xtab ka'anil,* "the cords of the sky," joined the intersection of the arches to four upright poles, beside each of which stood a man representing one of the sky-bearers. The complete structure was known as the sky-tree.[5]

SQUARING THE CIRCLE

The sky-tree altar of the contemporary Mayan shaman supplies us with more or less all the elements of a working model of creation. Thus, this may be a useful place to sum up what we have established so far about what such a model might consist of:

- the flat plane that is square and referred to as the "earth"
- the "wheel of the sky" itself, specifically the band of stars along the ecliptic; more generally, the whole sphere of the heavens, and referred to as "heaven" (or "sky")
- a central axis resting on the "primordial rock," and with the pole-star at its top, about which the circle of the stars turns
- the four pillars that mark the corners of the earth and act as its "foundations"
- "plumb-lines" or "guide ropes," suspended from the top of the central axis and determining the positions at which the four corners of the "earth" are "made fast" to the circle of the sky

The great mystery of creation is how the first two of these elements, heaven and earth, are to be conjoined. This is the mystery known as "squaring the circle." Its solution will produce a model of what the myths call the Created World.

The first stage in generating this solution requires the visualization of a system made up of a sphere with the globe of the earth at its center. What we have described as heaven, the circle of the stars, will appear on the inner surface of this sphere. At its top is the North Nail, the polestar, marking the summit of the axis of the world. What the base of this pole rests on is a different question; there are all sorts of answers, the "First Earth" being only one, and an unenlightening one at that. Another common answer is "a tortoise." As this is a mythical model, it all depends on why you want to know.

Any observer standing on the earth will see the pole of the axis at a height above the horizon that depends on the latitude of observation. Only at the North Pole will it be directly overhead. At the equator it will be on the horizon. Mythical traditions nevertheless persist in referring to the direction of this pole as "up" and its antithesis as "down."

Any Great Circle drawn through this pole will pass through the circle of the stars, "the wheel of heaven," at two opposite points. (For any sphere, a Great Circle is one with a diameter equal to that of the sphere. See fig. 5, page 49.)

A second Great Circle, passing through the poles at right angles to the first, will cross the equator of the sphere at two points midway between the first two, thus establishing four "corners." These two Great Circles make up the "girths" of the ship of creation, the plumb lines of the surveyed creation (fig. 7).

Now, it has presumably always been as clear as it is today that the "sky" does not remain fixed as viewed from the earth. While facing the night sky to the north (if we are north of the equator), we will always find ourselves facing the polestar; facing east or west will present differing stars, depending on the time of night or of year. To "make fast the sky" and complete the squaring of the circle must mean something more than merely holding heaven down to the earth with ropes and pegs.

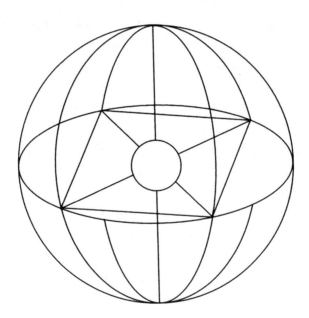

Fig. 7. The Frame of Time. The square earth, whose corners are marked by the intersections of the Great Circles and the celestial equator, the four pillars.

THE BIRTHDAY OF THE WORLD

So what kind of sense can be made of the idea of "making fast the earth to the sky"? First, it's worth stating once again that the earth whose creation we have been observing is not the earth on which we stand, nor is the sky the sky we see blue above us. We have seen that the earth is a plane through the equator of the sphere whose axis is marked by the polestar. The mythic "sky," in this instance, can be understood as the circle of the stars, composed of the groups of stars that stand along the sun's path, the ecliptic.

It is the establishment of the plane of the earth with its four corners fixed by the bonds of the colures* to the circle of the stars *at the correct points* that the myths describe as the surveying of the world, the imposing of order on chaos—in short, the act of creation.

A complete picture of the act of creation therefore demands the establishment of these correct points. We need to answer the question God asked of Job, "Whereupon are the foundations [of the earth] fastened?" Employing the particular logic of myth, the founding point can be said to be both that original stone set in the circle of eternity—that is, in the wheel of the stars—and the central point at which the axis rests, since both center and point of beginning, being bound together in a rigid framework, act as "foundation stones." We should not forget, however, that it is essentially the measurement of time that is being established, and that it is by using the sun as a measuring line that creation proceeds. It is the position of the sun that is the crucial fact in making fast the sky.

What we need to know is, where on the unending ring of the stars was the sun on the "birthday" of the world? In fact, this is not such an unanswerable question as it may seem. The birthday of the world is still widely celebrated. We call it New Year's Day.

As the year proceeds, the sun travels around the circle of the stars, and its position is deduced by observing the stars that rise immediately

---

*A *colure* is defined as a circle drawn on the celestial sphere passing through the poles.

before it. The sun is then said to "stand" in the stars that would normally follow but which are now invisible due to the presence of the sun itself. A full year's cycle will bring the sun around to its starting point. A New Year festival is therefore a celebration of the birthday of the world. It marks the return of the sun to the position it occupied not just at the beginning of the year, but at the beginning of time as well.

Such an idea is repeated in many traditions around the world, a point emphasized again by Eliade. (As a revealing detail, we might mention the Californian Indian New Year Ceremony, which includes the symbolic gesture of "putting posts under the world."[6]) In general, as Eliade says, "a periodic regeneration of time presupposes, in more or less explicit form, a new Creation, that is a repetition of the cosmogonic act."[7]

Unfortunately for our quest for the correct locations of the fixings of the sky, the date on which we celebrate the New Year in the West is almost totally severed from any relation to any defining cosmological points. It is a sign of how far removed from the meaning of time our sense of order has become. Fortunately, there are many records of the practices of extinct societies, and plenty of contemporary societies with more-direct links to the order of the cosmos. By observing their practices, we can begin to answer God's question.

It turns out that there are two very different ways of deciding on a starting point from which to measure the sky. The relationship between the two is crucial to the development of an understanding of the myths of creation. The first system is obvious to anyone watching the passage of the night sky. Or, at least, it would be obvious to an observer privileged enough to be able to enjoy the true glory of the spectacle. In our urban lives, we are almost completely deprived of what has been probably the most inspiring vision in the natural world, the Milky Way.

Step out into a frost-clear winter night unpolluted by streetlight or house light, and you cannot help but be struck by the grandeur of the sight. Arching above you across the sky flows the River of Heaven, as so many ancient cultures have named it. (In southern Africa it is called the Arch of the Queen of Heaven's Hut.) It crosses the path of the sun at two points, on more or less opposite sides of the sky, and offers the

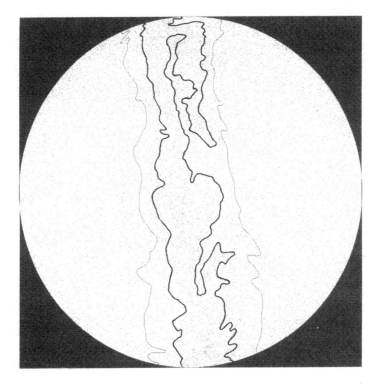

Fig. 8. The River of Heaven. The circle of the stars (south at bottom) divided by the Milky Way.

most immediate and practical "plumb line" for surveying the circle of the stars.

The Egyptians regarded it as the celestial Nile, the god Hapi, "Divider of Heaven," seeing the terrestrial Nile as dividing the land in the same way: "Hail Hap-ur [Great Hapi]. God of heaven, in thy name of Divider of Heaven."*

In her contribution to the book *Maya Cosmology*, Linda Schele explains how she deduced the fact that for the Maya, the Milky Way is the "World Tree." First Father "raised up the sky" by lifting this tree into its place, so that its crown stood in the north sky. She describes her own experience of viewing it: "We got out of the car and let our

---

*E. A. Wallis Budge, *The Book of the Dead*, 200, ch. 57. Faulkner has "O Hapi, Great One of the sky in thy name of 'the sky' is safe."

eyes adjust to the dark. There it was, arching across the sky from south to north—the great tree in a crystalline glory more powerful than anything I had visualised in my mind."[8]

By linking the ecliptic path of the sun to the northern sky, the Milky Way reveals one way to understand the mythical idea that central pillar and cornerstone are connected. There is much more to this image, however, and we shall have reason to return to it. In fact, it lies at the very root of creation history.

The second possibility for dividing the sky to measure its processes is more subtle. Nevertheless, it is the most widespread method used for defining creation, because it relates directly to more-pragmatic ways of measuring time. It derives from the fact that the sun does not rise and set at the same points throughout the year. In the Northern Hemisphere, in summer the sun rises and sets well to the north of east and west and stands high in the sky at midday, and the days are long. In winter, when the days are short, it rises and sets well to the south of east and west and reaches much lower in the sky at noon. This shifting of sunrise and sunset creates a "square" pattern whose "corners" mark the solstices of summer and winter, the points at which the sun reaches its extreme of rising and setting. In an echo of this pattern of the year, the Sudanese Dogon farmer "works in zigzag fashion, east to west, all the while progressing from north to south so that his work may go forward in the very steps of world movement."[9]

In a similar way, the Kogi people of northern Colombia see the pattern of the sun as a process of weaving: "Through the year, as the sun moves north, south and north again . . . it weaves the thread of life into an orderly fabric of existence."[10]*

This "square" pattern of solstice sunrises and sunsets, however, offers only a limited vision of the mythical square earth (although it is certainly an improvement on the view that regards the "corners" of the world as representing the compass points). It is a true square only at particular latitudes (roughly 55 degrees north or south). Elsewhere,

---

*For a fuller description of this image, see "The Temple," page 181.

it will generate a more or less elongated rectangle. What is more, such a vision retains the notion that the earth being created in the myths is the earth on which we walk, as defined by our own particular horizon. It does not of itself require any understanding of the contribution to be made by the stars.

To identify the determining square of the created earth, we need to return to the image of the two Great Circles from which the girths of creation are constructed (fig. 7). If one of these Great Circles passes through the two positions of the sun in the circle of the stars at the solstices, the other will pass through the sun's position at the equinoxes (the moments in the year when the sun rises due east and sets due west). These two circles are known as "colures," solstitial and equinoctial, respectively. It is these two colures that determine the four corners of the earth. Any one of these four corners might be chosen as the "beginning" of measurement, and examples can be found from around the world that use one or another.

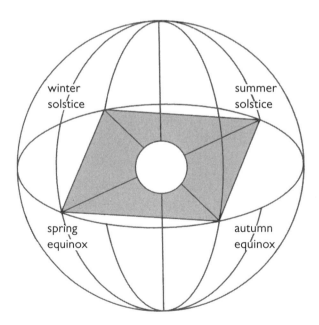

Fig. 9. The Ideal Creation. The square earth fastened to the circle of the stars by the two Great Circles, "the Bonds of Heaven," at the equinoctial and solstitial points.

We now have a picture of the ideal world as it was "first created." The two Great Circles of the colures, together with their common axis, bind the square earth into a rigid framework, fastened at the "correct" places to the wheel of the sky. This framework we might call the Frame of Time, and into it all the lesser degrees of order are woven. It serves as the supporting structure for the whole of the mythology of creation.

In a perfect creation, we would be able to say that the "square earth" is that flat plane that passes through the equator of the sphere and is defined by these four "corners." The whole significance of the myths of creation, however, is that the world is not perfect. The technique employed by the mythmakers to explain this fact is to suggest that there had once been such a perfect creation and that it came to an end. Ultimately, the dynamic between the ideal and the actual becomes the essence of the story.

We shall see that the truly perfect model of creation must in some way combine both methods of measuring the sky, using the equinoctial and solstitial colures in combination with the Milky Way. In fact, as we have seen, even the model we have built here of the square earth is itself an idealization and was never realized in actuality. Even this Frame of Time has about it more than a whiff of the mythical. But then, creation, as we well know, is far from perfect.

Before we can understand this problem, we must confront the question: If this is the created world, what kind of role is there for a creator?

The Myth of Corruption

THE GOLDEN RULE

FORGING THE SKY

In ancient Persia, the birthday of the world was celebrated in the first month of spring: "On the sixth day of Farwardin . . . is the Great Nauruz ['New Day'], for the Persians a feast of great importance. On this day—they say—God finished the creation."[1]

The festival itself, according to the great twelfth-century Persian epic poem *The Shahnameh*, was founded by the mythical ancestor of the Iranian race, Jamshid. This is the same Jamshid that Omar Khayyám describes in the Rubaiyat as being owner of the "Sev'n-ringed Cup," that is, of the seven planetary spheres of the cosmos.

Jamshid's epithet is Xsaeta, meaning "kingly," and this seems to be the origin of the Roman Saturn. Now, if we were to look for one name that might sum up the characteristics of the innumerable deities who have been credited with the work of creation by the varied races of humanity, it would have to be Saturn. Almost universally, in one guise or another, it is Saturn who is responsible for establishing the created world.

What makes a deity figure Saturnine, then? It would be a circular argument if we could point only to the act of creation to identify him. But it is the particular kind of creation that is the distinguishing factor. Not only is Saturn the first king; he is first and foremost a craftsman. The work of creation is frequently said to be the responsibility of a craftsman god (*deus faber*), who accomplished his work by dividing the eternal material and laying it out according to a plan that he had conceived, or had to hand. Moreover, the myths are generally agreed that the particular craft was that of the smith. It is as if it were out of iron that the craftsman built the houses of the gods, as was the case in Egypt and in Greece, where Hephaestus built the starry houses of the gods, or in Finland, where Ilmarrinen "hammered together the roof of the sky."

For as long as we retain the idea that mythical stories of creation are descriptions of the emergence of the physical world, ancient versions of the big bang, the manner of Saturnian creation, and the nature of Saturn himself are bound to elude us. Having established, however, that the events described in the cosmological myths are creation of a very particular kind and are in fact descriptions of the measurement of time on a cosmic scale, we are much better placed to understand Saturn's significance. However bucolic or sophisticated the imagery, the material is, in its origins, of the most cultivated nature. It deals with an advanced astronomical understanding and uses this understanding to describe the universe according to the laws of cosmic time.

From this point of view, we are perhaps less likely to be misled into thinking, when we read of Saturn as creator, that it is some mysterious or numinous deity. Traditions from around the world make it clear that it is first and foremost the *planet* Saturn that is responsible for the establishment of the ordered cosmos.

In the Iranian tradition, one version of the first age of the world is ruled by the Kaianian Dynasty. In this version it is the blacksmith Kavag who, using his smith's apron as a banner, usurps the monster who has taken possession of the throne. The word *kawi* in Persian means "smith" and is said to derive from this mythical smith-hero, but it seems to be directly related to the more ancient Hindu word for "poet or measurer" and to the Assyrian word *kawi,* which was both "smith" and the planet Saturn.

In Chinese tradition, "This planet was taken for the one who communicated motion to the universe and who was, so to speak, its king."[2]

Perhaps the most Saturnine of these creators is the Egyptian Ptah. He is credited with having founded the world, and his festival occurs once every thirty years, as did the Persian Kavian festival. This is the astrologers' "Saturn Year," the time taken by the planet Saturn to complete one revolution of the sky. For the ancients, this was the longest cycle of time to be readily discerned, Saturn being the farthest of the visible planets and therefore having the longest periodic cycle.

For these kinds of reasons, perhaps, Saturn is credited with being the "Father of Time" and the founder of the ordered universe. This subtlety has been appreciated by the better-informed commentators since the earliest times. The Roman Neoplatonist Macrobius makes it clear: *"Saturnus ipse, qui auctor est temporum"* (Saturn it is who is the author of time). Santillana and von Dechend quote an Orphic fragment that describes how Kronos, the Greek Saturn, "seems to have with him the highest causes of junctions and separations. . . . He has become the cause of the continuation of begetting and propagation and the head of the whole genus of Titans from whom originate the division of beings."[3]

THE LORD OF MEASURES

And how did Saturn/Kronos bring about this origination? Again in the words of Macrobius: "They say that Saturn cut off the private parts of his father Caelus [Ouranos]. . . . From this they conclude that, when there was chaos, no time existed, insofar as time is a fixed measure derived from the revolution of the sky. Time begins there."[4]

And the beginning of time is the beginning of measurement. The exact significance of this severing act of Saturn will concern us shortly. It is worth noticing here, however, that not only does Saturn generate time, but he retains the means to measure it as well. The Iranian deity Zurvan appears standing on the world egg holding the square and compass with which he will mark out the delineation of heaven and earth. His creative act is again one of "surveying." He lays down the measurements of creation and orders the construction in accordance with the kind of processes we have already seen attributed to the Hebrew Yahweh in the Book of Job.[5]

Around the world, one of the characteristics that distinguishes Saturnine figures is their access to or possession of the "Measures." The Mesopotamian Saturn, Ea, was known simply as "Lord of Measures."[6] These "Measures" are the treasure with which the world can be rejuvenated, as we shall uncover. In Saturn's possession, they are often referred to as "the Water of Life," but in whatever form, they appear to represent the power to define the Frame of Time. Santillana and von Dechend show how the planet Saturn might be understood to control the measures of time. They show the cycle of what are known as Great Conjunctions, the conjunctions of Jupiter and Saturn that occur approximately every twenty years, each time in a different part of the sky. Each successive third conjunction occurs in the sky slightly farther back than the first, so that an elaborate figure known as a trigon is produced that marks out a complete circle every 2,400 years. In this way, according to Orpheus, Kronos gives to Zeus "all the measures of the whole creation." (According to Plutarch's *The Face in the Moon,* Saturn, asleep in his Ogygian cave, is dreaming what Jupiter is planning.)

Even this challenging and demanding concept of the matter of time, however, does not describe the full nature of Saturn as creator. We have seen that the Frame of Time consists of both the celestial axis and the Great Circles fixed to the solstices and equinoxes. We should not be too surprised, therefore, to find that Saturn is also given such titles as "god of the center" by the Chinese, a term also used to describe the polestar. There is evidence that the Egyptians associated their Saturnine figures with the solstices, not to mention stories that relate Saturnine deities to the southern celestial pole. But more generically, the tradition is widespread that associates Saturn with the concept of kingship. Everywhere Saturn is the "First Emperor"—for the Greeks, South Americans, Mesopotamians, and Chinese. As Santillana and von Dechend note: "Even the Tahitian text 'Birth of the Heavenly Bodies' knows it: 'Saturn was king.'"[7]

Mircea Eliade sums up the widespread traditions concerning this ultimate creating divinity, from the "primitive" concepts found in Australia to those of the Arctic and central Asia, of Sumeria, India, and classical Europe. In all of these places the first god bears a name that means "sky" or "high," but Eliade stresses that this is not to be considered a

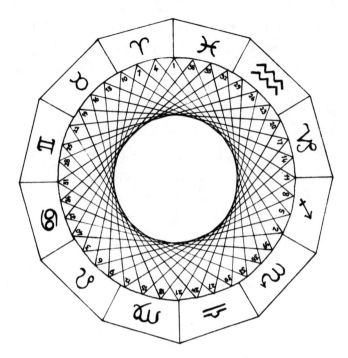

Fig. 10. Saturn the Measurer. The cycle of "Great Conjunctions" of Saturn and Jupiter. "The greatest Kronos is giving from above the principles of intelligibility to the Demiurge [Zeus]. . . . And Kronos seems to have with him the highest causes of junctions and separations" (Orphic fragment).

reference to "natural phenomena"; the notion is that the supreme god resides "above." In the Polynesian myth of Ta'aroa, the creator, having opened the "first shell," makes of it "the dome of the sky," the house of the gods. From a second shell, or perhaps from the bottom half of the first, he makes the "great foundation of the world."

According to Eliade, "It is the sky god that regulates the order of the cosmos, that dwells as supreme sovereign at the summit of heaven. He is the universal sovereign and his orders must be respected."[8]

Eliade goes on to point out that despite his sovereignty, this deity survives in many traditions only on the very periphery of religious life. He has no cults or worship, and many descriptions imply that he has withdrawn from his creation and has no concern for it. What is more, in many of these traditions he does not create directly but by means of an intermediary. This is the character that is sometimes described as the "Demiurge." This is the role that Saturn is seen playing: we can recognize him not only as the "sky" god, as monarch of the central pivot of creation, but also as that deity who has withdrawn and left others to manage. There are many stories of "First Emperors" who retire from their worlds to rest "in the depths"; King Arthur is only one of them.

We shall eventually be able to tie up these loose ends that seem to surround the creator deity; it cannot be overemphasized that the process of ordering chaos, of drawing a path across the trackless waste of nonexistence, is the process of measuring time. When Jamshid established the New Day festival, he did so to inaugurate a new era of time. It is Time that marks off creation from eternity and sets in motion the process whereby the eternal will manifest itself, or at least one pertinent aspect of the eternal will do so. The dilemma of the One God was that, in imposing order on the unmeasured, he not only revealed himself but removed himself from his creation as well. Being unable to exist in perfection within the limitations of the time-bound world, he had to leave a "mortal" deity to represent his immanence. As we shall see, such metaphysical meditations have their source in very real and observable phenomena.

For now we should return to the notion of the First Emperor, who is laying down the temporal measures for his perfect creation. It is common for his story to masquerade as, or intrude into, what might be considered to be history. Nowhere is this more the case than with histories that purport to tell of the origins of dynasties. The "First Emperor" of Chinese tradition was Huang-Ti, who "established everywhere the order for the sun, the moon and the stars."[9] (Huang-Ti was an "unmistakable smith," according to Santillana, and is "acknowledged to be Saturn.") But all the ancient Chinese emperors were regarded as "rulers"; they carried, as an emblem of their role, a measuring rod with which they could "call down the right numbers of Heaven": "The ruler maintains the virtue of order of his realm by receiving through inspiration from above the right numbers of the heavenly calendar. Similarly, a decadent virtue can draw down the cosmos into disorder by bringing the numbers of the calendar out of their order."[10]

DREAMTIME

The world that Saturn created was not the world in which we live; it is regarded the world over as being quite distinct from all the numerous worlds that have succeeded it. The "Saturnae regna," the Age of Saturn, was everywhere considered to have been a time of perfection. For the ancient Iranians, the first king was Yima. According to the earliest tradition, he reigned for a thousand years: "In his reign there was neither cold nor heat, old age nor death nor yet disease."[11]

This was truly the Golden Age. It is universally said to have been one without death, conflict, or corruption. The Australian Aborigines call this era the Dreamtime. According to Eliade, "[T]he Australians consider that their mythical ancestors lived during a Golden Age in an earthly paradise in which game abounded and the notions of good and evil were practically unknown. It is this paradisal world that the Australians attempt to re-actualize during certain festivals."[12]

For the Egyptians, their "First Time" (Zep Tepi) was a time of perfect peace, an age when the skies had a magnificent balance. It is this notion of a celestial balance that must lie at the heart of the myths of Saturn's perfect reign. From many traditions we learn that at that time, men and gods walked and talked together. Not only were heaven and earth arranged so that they were in perfect harmony; there was, in fact, a direct link between them. The Book of Genesis describes "Yahweh walking in the garden in the cool of the day,"[13] and West African traditions echo this idea: "Long ago men were happy, for God dwelt with them and talked with them face to face."[14]

Having established a reliable way to interpret the meaning of these ubiquitous and ambiguous words *heaven* and *earth,* we are in a position to make a new approach to the problem of what such a perfectly balanced arrangement might imply. It may also be possible to say something about the kind of link that once existed between the two, for the story of separation tells us that this link was broken when heaven and earth were separated, so that men and gods no longer walked together, and it was from this severance that all the aspirations of the human spirit ensued.

We have established in previous chapters that the "square earth" is the figure delineated by the four corners of the solstices and equinoxes; that the "circular heaven" is the path of the planets through the ecliptic; and also that this circle is divided by the Milky Way. It is clear that a harmonious relationship between heaven and earth would require these two systems of measurement to coincide in the most satisfying way. One or another pair of the four corners must coincide with the crossing points of Milky Way and ecliptic, in one of the two possible arrangements. Before we can say which of these arrangements marked the actual era of the beginning of time, the moment of creation, we must understand more about how the First Time came to an end.

Whether the Golden Age is inhabited by the Children of Heaven as it was in Egypt or by Adam and Eve in Eden, the fate of this perfect world was to be the same. As a result of some transgression, the occupants were to be driven out, exiled from their gods, and wander the time-bound world searching for the way back to that original harmony. Eliade recounts this story as it is told by the Dayak people of northern Borneo: "The first human beings lived in a time of peace and plenty. (They were living at that time in the mouth of the encircling watersnake.) But the stories tell how, because of their transgressions, either their descendants had to leave and set out on their journey into Time, or this perfect world came to an end, leaving only the world as we know it."[15]

This story is invaluable, not only because it forms just a small part of wonderfully rich tradition of cosmological thought from one of the most remote corners of the world (a tradition that, incidentally, relates intimately with those from more familiar parts), but also because it makes it explicit that it is a journey into time that begins when this perfect age comes to an end. Paradoxically, Saturn's Golden Age is not only the beginning of time, but it is also timeless. The "First Time" was not subject to time as all subsequent eras have experienced it, our own included. This paradox can be resolved, but not until we have a clearer picture of what caused the demise of this perfect world.

PRIDE AND FALL

OUT OF THE GARDEN

Few stories can be more familiar to us in the West than that of Adam and Eve and their exile from the Garden of Eden. It may come as a surprise, however, to discover that the same story is told in many other parts of the world. The First People, whoever they may be, having originated in the perfect world where evil and death are unknown, are obliged, usually as a result of their own transgressions, to depart and begin a life of toil in the time-bound world, enslaved to death and disease as well as the inevitable inroads of corruption.

As for the cause of this exile, the exact nature of the transgressions of these first people, Eliade has this to say: "In most of the Pacific myths, the catastrophe is caused by some ritual misdemeanour."[1]*

This kind of fault results from ignoring the dictates of what the

---

*The same idea was recorded from African tradition: "Mulungu in the beginning lived on earth, but went up into the sky because men had taken to setting the bush on fire and killing 'his people' [i.e., the animals]. The same or a similar idea (that God ceased to dwell on earth because of men's misconduct) is found to be held by other Bantu-speaking tribes, and also by the Ashanti people in West Africa and the 'Hamitic' Masai in the east." (Alice Werner, *Myths and Legends of the Bantu,* 1911; www.sacred-texts.com/afr/mlb/index.htm.) Werner goes on to add, "It may be connected with the older and cruder notion (still to be traced here and there) that the sky and the earth, which between them produced all living things, were once in contact, and only became separated later."

Hindu lore calls Rita. When Varuna "made fast this aweful heaven and earth," he is said to have done so by fastening the sky "to the seat of Rita." Eliade explains that Rita "designates the order of the world—an order that is at once cosmic, liturgical and moral. The creation is proclaimed to have been effected in conformity with Rita; it is repeatedly said that the gods act according to Rita, that Rita rules both the cosmic rhythms and moral conduct. . . . The seat of Rita is in the highest sky."

This, it seems, can be little less than the law of the universe. Similar concepts existed across the world. According to Stechinni, the Egyptians used the word Ma'at or Maet to describe the same construct, "a word which we translate as 'truth' or 'justice,' but which has the extended meaning of the proper cosmic order at the time of its establishment by the Creator. For it was believed that the gods had first ruled Egypt after creating it perfect."[2]

We might just call this principle "God's commandments" (except for the fact that when personified, this concept is most frequently female), as long as we do not forget what we have been at pains to establish—that "God's creation" was made by working with the "unwrought matter of Time," and that God's commandments are expressed in the language of time.

The proper role of both kings and people, whether they are pharaohs or First People, is the correct observance of ritual (the word itself seems derived from the Sanskrit *rita*). That is to say, they must reproduce faithfully on earth the practices of the "gods" in heaven. If "sacrifices" begin to be held in a disorderly manner, it is as a result of the ignorance, or the ignoring, of the commandments. The emperor, by his decadent virtue, can draw down the cosmos into disorder by "bringing the numbers of the calendar out of their order."[3]

In reading these tales of the demise of the perfect world, it is sometimes easy to lose contact with exactly which period of mythical time is involved. The creation and its people seem fated to repeated destruction. According to many widespread accounts, they periodically disappear in the Deluge as a result of their sins. On all occasions, however, this word *sin* is to be understood in the sense in which it is used in the Hebrew scriptures, to mean exactly this kind of "ritual misdemeanour."

Occasionally it is mere overpopulation on earth that drives the gods to destruction, but as we shall see, it is precisely "overpopulation" that can lead to disorder. The important point to recognize is that even if the First People are punished for their wrongdoing, and even if, as appears to be the case with Adam and Eve, their particular transgressions are those of pride rather than ignorance, they are ultimately only reflections of much grander goings-on in the heavens above them. If people fail to match up to the dictates of the gods, it is because the gods themselves are falling short of the mark.

CORRUPTION IN HEAVEN

The story of the "transgressions" of the First People is retold in greater detail in the Rig Veda traditions that describe Varuna as the one who props apart heaven and earth (the verb used for "prop" is *skamb*, that is to say, it is the action of erecting the world-scaffold). He is described as being one of the Asuras, the deities who ruled the first age of the world but whose encroaching corruption led to its downfall: "Assuredly were the Asura originally just, good and charitable, knew the Dharma and sacrificed . . . but afterwards as they multiplied in number, they became proud, vain, quarrelsome. . . . They made confusion in everything. Thereupon, in the course of time . . ."[4] Their age was doomed to perish.

Clearly, these Asuras are not members of humankind as we understand it, although they too appear to be bound by the laws of Rita. Indeed, the Asuras of the Rig Veda are gods themselves—the first gods, that is. Ultimately, the order of beings described by the Egyptians as the "Divine Children" are responsible for the downfall of the perfect world, as we learn from the events described in chapter 175 of the *Book of the Dead*: "What is it that hath happened unto the divine children of Nut [the goddess of the over-arching sky]? They have done battle, they have upheld strife, they have done evil, they have created the fiends."[5]

The suggestion seems to be that these "First People," those whom we might call the Ancestors, who inhabited Saturn's perfect world, were not people in the way we are, just as their world is so unlike ours. They appear to occupy some position between the Godhead and humanity, and are often described as being the "offspring of the primordial heaven and earth."

These offspring come to life initially entrapped in the eternal congress of their parents. As they begin to multiply, there is eventually insufficient room for them all. The Maori describe the situation:

> According to the traditions of our race, Rangi and Papa, Heaven and Earth, were the source from which in the beginning all things originated. Darkness then rested upon the heaven and upon the earth, and they both clove together, for they had not yet been rent apart; and the children they had begotten were ever thinking amongst themselves what might be the difference between darkness and light; they knew that beings had multiplied and increased, and yet light had never broken upon them, but it ever continued dark.
>
> Rangi, the Sky Father, felt love for Papa-tu-a-nuku ["The Earth"], who lay beneath him, so he came down to Papa. At that time absolute and complete darkness prevailed; there was no sun, no moon, no stars, no clouds, no light, no mist—no ripples stirred the surface of ocean; no breath of air, a complete and absolute stillness. . . . Then Rangi clave unto Papa, the Earth Mother, and held her close in his embrace, and as he lay thus prone upon Papa, all his offspring of gods which were born to him, both great and small, were prisoned beneath his mighty form and lived cramped and herded together in darkness.[6]

This may not sound like a perfect world. For elucidation, we can read in greater detail about the existence of similar beings elsewhere. The Babylonian creation epic Enuma Elish, for instance, tells how the first couple, Apsu and Tiamat, whose "mingled waters" were the primal manifestation, brought forth gods, although they remained at that time unseparated. The epic gives us their names: "Lahmu and Lahamu were brought forth, by name they were called. For aeons they grew in stature. Anshar and Kishar were formed, surpassing the others. They prolonged their days, added on the years."[7]

Within this unseparated world, the children and grandchildren of the primal couple live the life of heroes of "broad wisdom, understanding, mighty in strength." But their relentless multiplication eventually becomes intolerable for their parents, whose anguish may sound familiar: "The divine brothers bonded together; they disturbed Tiamat as they surged back and forth; Yes, they troubled the mood of Tiamat by their hilarity in the Abode of heaven. . . . Unsavoury were their ways, they were overbearing."[8]

The overbearing behavior of these divine brothers reminds us of the confusion caused by the growing pride of the Asuras. Unlike the Asuras, however, the Babylonian Divine Brothers devise a plan to escape from their parental restrictions.* Although at first sight this story may seem to differ in its significance from those that tell of the demise of the perfect world, the final outcome is the same: the severance of the primal couple and the establishment of a new regime of order. What is more, this is an order quite different from that which has gone before; it must defend itself relentlessly against the inroads of the forces of chaos, which are threatening perpetually to engulf it.

We might well say that it is the recognition of this ever-present threat of destruction by the forces of chaos that distinguishes all subsequent creations from that first and perfect order. What connects these forces with the "transgressions" of the First People is the notion of sin, particularly the sin of pride.

Perhaps the most striking example of this destructive act of pride is that told in the book of Isaiah 14:12–15 concerning Lucifer: "How art thou fallen from heaven, O Lucifer, son of the morning! For thou hast said in thine heart, 'I will ascend into heaven, I will exalt my throne

---

*In the myths of the Pacific Islanders, the plan is prepared with the cooperation of the primordial parents (see the appendix).

above the stars of God: I will sit also upon the mount of the congregation, in the sides of the north [where the City of God is on Mount Zion; see Psalm 48:2]: I will be like the most high.' Yet thou shalt be brought down to hell, to the sides of the pit."[9]

Without this act of consummate pride, there could have been no expulsion from Eden. But it was not just an act of individual will. Something had gone wrong with the whole host of heaven, as the text of Psalm 82 suggests when Yahweh says to the assembled host, "I say you are gods, sons of the Most High, all of you, nevertheless you shall die like men and fall like any prince."[10]

No wonder ordinary mortals were struggling. Again, in Psalm 58, Yahweh asks, "Do you decree what is right, you gods?"[11]

Now, if the gods start issuing decrees that are not right, what hope is there for the maintenance of order? But we can glean some valuable information from these verses regarding the "host of heaven," who, in other parts of the Hebrew scriptures, are called "the Sons of Heaven" or the "assembly of Holy Ones," or, more simply, "the Sons of God." The commentaries to the Jerusalem Bible translation make it clear that, for the early church fathers, at least, these were the angels—that is to say, the stars.

These are obscure areas of biblical tradition, but there are also stories that suggest that the work of creation itself was accomplished by a deity that we might consider to be "evil." Such was the opinion of various sects that flourished around the time of Christ, whose only goal was to remove themselves from "the weary wheel" as soon as possible, to avoid at all costs introducing further lives to the torment by remaining celibate, and, while they were obliged to exist in it, generally to have as little to do with creation as possible. According to the Gnostic tradition as transmitted by Valentinus, the creator Yaldabaoth himself had been the perpetrator of this overbearing pride, claiming that he was the highest god and that none had been before him. (As a result, Wisdom appears and throws him from his place.)

This is an extreme version of the more general opinion that it is as a result of some act of "pride" that humankind and the whole of creation are in bondage and destined to perish. The psychologist Joseph Campbell ascribed this act of pride to the tyrant "Holdfast": "an impulse to egocentric self-aggrandisement." He saw it as the obstinate resistance to the inevitability of change. "And so," says Campbell, "the One was broken into the many, and these then battled against each other."[12]

Behind all these ideas lies the dilemma that Plato encapsulated in his description of the craftsman-god faced with the brute matter of time, which was unable of itself to manifest a perfect creation. This is a profound notion that we shall have to return to. It would seem that even the Golden Age had been subject to time. However close the world of the Divine Children in Egypt may have seemed to the eternal, it had known some kind of time, since it was their particular crime to disrupt it; "they bring their years to confusion and they throng in and push to disturb their months; for in all that they have done unto thee they have worked iniquity in secret."[13]*

That is to say, the perfect harmony of the first creation was perfect in appearance only; given enough time, it was bound to fall prey to the monster of chaos.

BEYOND MEASURE

Once the "Divine Brothers" of Babylonian tradition have disposed of their grandfather Apsu (the Depths), they are soon confronted with their grandmother Tiamat in her embodiment of the monster of chaos. She brings with her a host (eleven, in fact) of misshapen monsters that she has bred. These monsters have the head of one animal and the body of another. Together with their leader, Kingu (Tiamat's "first-born," whom she has elevated to be her "only consort"), they seem to represent what Jung pictured as "the unfettered opposites in Chaos," that is to say, the confused disarray of the zodiacal signs. The gods, however, elect

---

*Faulkner treats this passage from the Egyptian Book of the Dead as an injunction to Thoth to "shorten their years, cut short their months because they have done hidden damage to all that you have made."

Marduk to defend them, and he promptly slices Tiamat in two and puts the rest to flight.

Marduk then uses Tiamat's body as the structure for a new creation: "He crossed the heavens and surveyed the regions; He prepared the mansions of the great gods, he fixed the stars to correspond to them; He ordained the year, appointing the heavenly figures; for each of the twelve months he fixed three stars."[14]

Such is the nature of mythical creation. And the threat that all the varied "monsters of chaos" pose is that they will upturn this celestial order. In Iran, the monster Azhi, whom Yima (Saturn) at the beginning had bound at the bottom of the ocean, rises up and saws him in two, usurping his authority (which Yima had prejudiced as a result of his proud boasting). When this monster is eventually destroyed, he is then thrown out of the sky, "by the same hole through which he had first rushed in." The location of this hole is said to be at the "meeting place of this contaminated world of time and space and the pure uncontaminated world of eternal truth."[15]

Fig. 11. The Great Cat, Re, destroying the serpent Apep "on that night of making war, when the Children of Impotence . . . entered into the east of the sky and war broke out in the entire sky and earth." Apep (Apophis) is the Egyptian serpent of primeval chaos. From the Egyptian Book of the Dead, the papyrus of Hu-nefer.

What is rushing in at this mysterious point is the disordered time-lessness that predates the establishment of measurement on earth. To the Zoroastrians, the monster threatening creation was known as Az, a name that is sometimes translated as "concupiscence, lust, greed." In his work *The Dawn and Twilight of Zoroastrianism*, R. C. Zaehner tells us more about this creature: "Her essential activity is disorderly motion or disruption. . . . Az is the principle of disorder that has invaded the natural order . . . [the] ignorance of the right order of things."[16] What was held to destroy the world, says Zaehner, was "deviation from the Aristotelian mean which the Zoroastrians interpreted as meaning the orderly arrangement of a cosmos created by God."[17]

The ancient Greeks regarded this orderly arrangement to be main-tained by the principle of "measure," in which, according to the physicist David Bohm, "measure was not looked on as . . . some sort of compar-ison with an external unit; this latter was regarded as a display of an inner measure, which played an essential role in everything. Something which went beyond its proper measure was inwardly out of harmony, so that it was bound to lose its integrity and break up."[18]

This "going beyond the proper measure" lies at the heart of all these stories about pride and corruption, about transgression and sin. When the stars themselves, the Sons of the Most High, give decrees that are not right, the Chinese emperor can do nothing but "bring down the numbers of the calendar out of their order." As a result of his "decadent virtue," itself only a manifestation of the decadence of the heavenly host, the links between heaven and earth must be broken, and his once perfect reign must come to an end.

The words used for *sin* in the Hebrew scriptures all mean something like "to fall short of the mark."* To understand the nature of the origi-nal sin, we must appreciate that the mark in question is that which was laid down upon undifferentiated chaos at the creation. It is the mark of

---

*The *Encyclopaedia Judaica* says, "In Biblical Hebrew there are about twenty different words which denote 'sin'; *het* ('to miss, fall'), *pasha* ('to breach a covenant'), and *awon* ('to become bent') are the most common."

time, whereby cosmic order was created. The disorderly motion that the serpent Az willfully introduces into creation results in the obscuring of this mark. It symbolizes the fact that something is seriously wrong with the system by which time itself is regulated. The Divine Children are bringing "their years to confusion" with disastrous results. In the end, it is time itself that will wreak vengeance. In the Bhagavad Gita, Vishnu reveals his nature as creator and destroyer of the world: "Know that I am Time, that makes the worlds to perish when ripe and bring on their destruction."[19]

THE SEPARATION
OF HEAVEN AND EARTH

So when the last hour shall so many Ages end,
and thus disjointed, All to chaos back return,
then all the stars shall be blended together,
those burning lights on high in sea shall drench
Earth then her shores shall not extend, but to the waves
    give way,
the moon her course shall bend cross to her brother's and
disdaining still to drive her chariot wheel athwart the
    heavenly orb,
shall strive to rule the day.
This frame to discord bent, the world's peace shall disturb,
and all in sunder rent.

LUCAN, THE CIVIL WAR, BOOK I, 83–92

AT THE WORLD'S END

Once the Children of Heaven have begun their iniquitous workings on
the machinery of time, the world is set upon a path toward destruction.
The relentless pushing and straining will increase until the whole frame
of the earth parts company with the heavens. What causes the destruction of the world is the undermining of the cosmic law, the Hindu Rta,
the Egyptian Ma'at. As a result, the integrity of the time-bound world

is being threatened and ultimately overrun by the waters of eternity. All those undivided aspects of potentiality that have lain dormant during the first age of the world are now said to be awakening and attempting to impose themselves on time and space.

Pride, we should remember, comes before a fall. The failure of creation to recognize the dictates of the law has stretched beyond endurance the bonds that once fastened the sky and held back the waters. Now creation shakes itself to pieces as the forces of order struggle vainly against those of chaos, leading to an almighty catastrophe that shatters the whole structure. Foxes or wolves gnaw through the ropes that bind the universe together; trees that support the sky are felled or dug up; stones whose job it is to restrain the overpowering forces of chaos are removed from their positions, enabling the waters of the abyss to engulf the decaying order; pins and nails that should be holding the universe together are removed or shot down; the heavens fall, setting fire to the earth; huge dragons and other beasts appear on the earth to engulf humankind and destroy the world.

The supply of variations seems inexhaustible. The mythmakers employ all the resources at their command as, with obvious delight, they begin to tell of the "Mother of Battles." Not only are there the massive flights of fancy that openly describe the end of the world as a battle between the various gods, such as the Götterdämmerung or the numerous battles described in various Indian scriptures, but there are also many tales that profess to be concerned with other, sometimes more historical events that can in fact trace their material and their structure back to this basic story.

One such is the story of the battle of the Plain of Tuired, which occurred in the mythical history of Ireland. It is fought between the gods of Danu and the "People of the Sea," the Fomori, strange, misshapen creatures who have for their leader the mysterious Balor, the "god of the underworld." Balor is ultimately defeated by Lug, who is chosen as champion by the terrified gods, just as Marduk was chosen in Babylon and Indra in India.

The Welsh story of the Battle of the Trees, described in detail by

Robert Graves in his book *The White Goddess,*[1] is another version of the end of the world, disguised as a description of something quite different. Similarly, the book of Isaiah tells of the fall of the House of David in totally cosmological terms: "In that day shall the nail that is fastened in the sure place be removed and cut down and fall and the burden that was upon it be cut off . . . the windows from on high are opened and the fountains of the earth do shake . . . the earth is utterly broken down . . . then the moon shall be confused and the sun ashamed."[2]*

This removal of the North Nail, or of the axle of the Cosmic Mill (or of one of the stars of the Pleiades, as in Genesis), allows the "fountains" to open, and the Flood pours down to engulf creation.

That is to say, to engulf the *current* creation, for however final these momentous battles seem, there is nevertheless the feeling that it has all happened somewhere before. Something always remains of the former existence, the residue of creation, from which a new world will emerge. We have already seen how the eternal Atum in Egypt said of himself, "When earth will become again part of the Primeval Ocean, Then I will be what will remain,"[3] and how in India it is the serpent Sesa, whose name means "Remainder," who, after the destroying the world with his fiery breath, lies at the bottom of the abyssal waters, waiting for the time to come when he and Vishnu will continue the work of creation.

Indeed, the Indian stories include so many episodes of cosmic destruction in so many differing formats that it is impossible to tell which occasion is being described in any one text. Elsewhere, the destruction of the world is more singular, but nonetheless temporary. In the Norse myth of Ragnarök, the "Twilight of the Gods," at the close of the battle, the Sybil describes how the daughter of the sun will survive the universal destruction and how a new world of gods will set about reestablishing the earth: "Now do I see the earth anew Rise all green

---

*Isaiah 24:4, 5 explains that this is a consequence of the behavior of "the haughty people of the earth . . . they have transgressed the laws, changed the ordinances, broken the everlasting covenant."

from the waves again. . . . The gods in Idavall meet together, Of the terrible girdler of earth they tell."[4]

A similar revelation was made to St. John the Divine: "And I saw a new heaven and a new earth; for the first heaven and the first earth were passed away."[5]

That the world has its source in the eternal and will one day return to it is an ancient and widespread idea. It is almost invariably accompanied by the idea that this is a continuous process, worlds coming into being and returning to their source in an endless succession. Thus, the creation myth as told in Hawaii begins with the origin of a new world from the shadowy reflex of one that is past: "Although we have the source of creation from chaos, it is a chaos which is simply the wreck and ruin of an earlier world."[6]

Anaximander, writing in Greece in the sixth century BCE, summed up this notion: "The first principle of existing things is the boundless for from this all things come into being and into it all perish. Wherefore innumerable worlds are brought to birth and again dissolved into that out of which they came."[7]

Eliade attempted to explain this concept: "By the mere fact that it *is,* that it lives and produces, the cosmos gradually deteriorates by falling into decay."[8] His argument is really circular, though, for the creation as we have described it can only be said to "live" precisely because it is said to die. To understand why the created worlds must come and go in this way requires one final element in our model.

LOST STARS

What is it, then, this force that makes the framework of creation shake itself to pieces? If the universe must inevitably end, what kind of understanding of this process do the prophets have that gives them confidence in their promise of a new world to come? The answer to this question is the ultimate secret of the ancient wisdom. It is hardly surprising that it can nowhere be found except in veiled and cryptic forms. To begin with, we can turn to the apocryphal Book of Enoch, where, in chapter

18, verses 11 through 16, Enoch is describing the vision he has received, in which he is guided by the angel Uriel through the mysteries of the heavens:

> And I saw a deep abyss, with columns of heavenly fire, and among them I saw columns of fire fall, which were beyond measure alike towards the height and towards the depth.
>
> And beyond that abyss I saw a place which had no firmament of the heaven above, and no firmly founded earth beneath it: there was no water upon it, and no birds, but it was a waste and horrible place.
>
> I saw there seven stars like great burning mountains, and to me, when I inquired regarding them, the angel said: "This place is the end of heaven and earth: this has become a prison for the stars and the host of heaven. And the stars which roll over the fire are they which have transgressed the commandment of the Lord in the beginning of their rising, because they did not come forth at their appointed times. And He was wroth with them, and bound them till the time when their guilt should be consummated."⁹

The cause of the destruction of the world, then, is not the transgressions of humankind, nor of the First People, nor even, on the face of it, of the Divine Children, but rather of the stars, "for they came not in their proper season." What is recorded in these stories of destruction, of the breakdown in cosmic order, what has determined all this mythologizing, is the knowledge that the stars do not forever revolve around the same point in the sky.

We look to the north and expect to see the polestar, which we call Polaris, fixed at the center of the circling stars, like Prince Dhruva, around whom the heavens turn. But it was not always so and will not be so again in ages to come. Certain other stars, now circling the pole, will in their turn become the pole themselves, and the heavens will revolve about them. It was just such a replacement that had caused Prince Dhruva to be appointed to his elevated position.

And not only the pole will change, but the whole frame will shift as

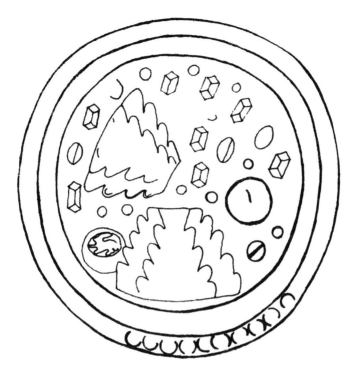

Fig. 12. The Broken Mountain. The collapse of Mount Meru, the hourglass mountain at the center of the world, home of the gods. Sun and moon (with its hare) are dislodged and stars (?) fall from the sky.

well, causing all the stars that mark out the pillars of the ordered world, the corners of the earth, to be removed from their positions. Those deities whose names are the titles of these pillars will be deposed and new deities inaugurated. Hence in India, we are introduced to "new Indras" and "new Agnis" as the old ones resign or are displaced. It is just such events that the Purana scriptures record as the consequence of the triumph of the buffalo demon:

> The general of the demons and the other demon chiefs . . . said to the
> buffalo demon, "Formerly we were kings in heaven, O clever one, but
> our kingdom was forcibly stolen by the gods. . . . Bring that kingdom
> back to us by force." . . . A fierce hair-raising battle between the gods
> and the demons then took place for a hundred years, and at first the

multitude of gods was put to flight in all directions. . . . They went to Brahma in terror . . . and he reported to the two gods Sambhu and Krsna what the buffalo demon had done: "He has thrown Indra, Agni, Yama, the sun, the moon, Kubera, Varuna and the others out of their positions of authority. . . . He has brought all the elephants of the regions of the sky . . . to live in his own palace. He has struck the ocean with his horn."[10]*

The great hero was angry . . . and tossed the mountains high with his two horns. . . . The ocean lashed by his tail . . . overflowed on all sides; the clouds, pierced by his swaying horns, were broken into fragments; and mountains fell from the sky by the hundreds.[11]

THE INCLINED EARTH

In the telling of similar events in China, the full import of this displacement appears:

Once upon a time a horned monster, Kung Kung, ventured to fight one of the five sovereigns for the title of emperor. He was overcome and in his rage he flung himself at Mount Pu Chou ["broken, non-circular" . . . the "cosmic mountain of the North West"], or more literally impaled it on his horns. The column of the sky was broken, the link with earth was cut. In the north-west the sky collapsed. Hence the sun, moon and stars slipped towards the north-west and the earth tilted to the south-east. Thereupon the waters spread and flowed to the south-east and flooded the empire.[12]

Sky and earth have parted company, the sky collapsing and the sun and stars slipping to the northwest, while the earth has tilted to the southeast. These events may sound strange, and an insubstantial argu-

---

*According to ancient Indian cosmology, eight celestial elephants protect the eight points of the compass or regions of the sky.

ment for an explanation of the destruction of the entire cosmos, but they reappear consistently in many remote places. The Book of Enoch records that "in those days Noah saw that the earth became inclined, and that destruction approached."[13]

Eliade reports that among the Inuit peoples of the far north, "according to a fairly general belief, the world tilted at one time, and its former occupants now live underneath"; and a Cree Indian elder, speaking of the Medicine Wheels, said, "My old people said these were before our time. There was another creation, then a shuffling of this earth, you might say, then we were created."[14]

These stories, though minimal in themselves and buried in the remote backwaters of traditions with which we are far from familiar, tell of major events in often surprisingly explicit ways. More-familiar tellings, however, speak less precisely, but more grandiosely, of the separation of heaven and earth.

We have already seen how the two primordial parents of the world had once lain in "eternal congress." In the Babylonian story, Apsu finally decides to deal with his unruly offspring: "I will wreck their ways/ That quiet may be restored,"[15] he says. But one of the divine brothers, Ea, overhears the plan and decides to act quickly: "Ea the all-wise saw through their scheme. Having fettered Apsu, he slew him."[16]

Ea then goes on to establish the first creation, wherein he is known as "the Lord of Measures." This first lord is the Sumerian version of the Greek Kronos, whose parents, Ouranus and Gaia, were oppressing their children, the Titans, by remaining in continual congress. Kronos is the one who opts to separate the primordial pair, using a curved knife supplied by his mother, who is said to have resented her husband's oppression. The meaning of the act of castration is obscure; that more than mere severing is involved is suggested not only by the emergence of Venus from the foam produced by the "private parts" of Caelus (the Roman Ouranus) when they fell into the ocean but also by a story from the Yoruba people of West Africa, who tell how "Ifa took the materials of the universe from a snail shell in a bag suspended between the legs of an older god and used them to form the universe."[17]

Fig. 13. Lifting the Heavens. The Goddess Nut (the Sky) is held above Geb (the Earth) by their son Shu ("Air").

In India, Dyaus and Prithuvi are separated by Indra, according to the Aitareya Brahmana. There are also suggestions of the union and divorce of the pair in China. In Egypt it is Shu ("Air") who separates the divine parents Nut and Geb.

Thus are heaven and earth once and for all parted. The links with the gods are severed, and light and mortality enter the world. In short, it is the separation of the primordial parents of the world that produces time, as Macrobius explained: "They say that Saturn cut off the private parts of his father Caelus. From this they conclude that when there was chaos, no time existed, insofar as time is a fixed measure derived from the revolution of the sky. . . . Time begins there."[18]

Macrobius's remark concerning the direct relationship between castration and creation of time appears to be a classic non sequitur, but in relating the beginning of time to the severance of heaven and earth, whereby the "revolutions of the sky" can commence, he is alluding to the secret knowledge that all these myths record, the ancient knowledge that the axis of the earth and that of the heavens are not the same. As a result of this phenomenon, the celestial equator, that plane which we have called the mythical earth, which is at right angles to the axis of the earth's rotation, is at an oblique angle (of about 23.5 degrees) to the circle of the stars that marks the planetary paths (the plane of the ecliptic) that we have called the mythical heaven.

It is this fact that results in the existence of time, on a cosmic scale, and that forms the foundation of all the storytelling concerning the creation and destruction of the worlds. Milton describes the establishment of this tilt, known prosaically to astronomers as the "obliquity of the ecliptic," as God's response to the disobedience of Adam and Eve:

> *Some say he bid his Angels turn askance*
> *The poles of Earth, twice ten degrees and more*
> *From the sun's axle, they with labour pushed*
> *Oblique the centric Globe:*
> *Some say the Sun*
> *Was bid turn reins from the equinoctial road*
> *Like distance breadth to Taurus with the seven*
> *Atlantic Sisters and the Spartan twins*
> *Up to the Tropic Crab, thence down again*
> *By Leo, and the Virgin, and the Scales, as deep as Capricorn,*
> *   to bring change*
> *Of seasons to each clime: else had the Spring Perpetual*
> *   smiled on Earth.*[19]

As Milton points out, were this not the case, were the earth and the heaven still "in eternal congress," unseparated, then perpetual spring would smile and conditions on earth would be those of a constant equinox, with day and night of equal length and no seasonal variation. Milton suggests that this had been the case prior to the expulsion of Adam and Eve from Eden. It was their disobedience that caused God to order the end of this "timelessness."

The impact on the ordered universe of the obliquity, however, is more dramatic than this, and it is time on a scale much grander than the merely seasonal that Saturn originates. As a result of the separation of heaven and earth, the system of measurement defined by the position of the stars in relation to the sun is not reliable and stable; the seasonal cycle of the sun that establishes the frame of time does not remain fixed to the circle of stars.

This is why the model of the Frame of Time proposed in earlier sections is inadequate. The square earth described there has become "inclined" and no longer lines up with the plane of the celestial equator. Indeed, this is the very meaning of the separation of heaven and earth. As a result of this separation, over long periods, the gods become removed from their positions of authority. The stars, the Sons of God, as the Hebrew scriptures call them, transgress the commandments of God and fail to come in their proper time. They no longer rise in a way that enables the demarcation of order. The Sons of the Most High begin to issue decrees that are not right, the emperor calls down the numbers of heaven out of order, and his position of authority becomes as shaky as that of the ruling stars he represents. The vast cycle of time lurches ever forward.

NIGHTS IN
THE GARDEN OF EDEN

THE WHEEL OF TRANSCENDENT REALITY

When Saturn separated the starry circle of heaven from the seasonal solar year of the square earth, he set in motion the machinery of time on its grandest scale. This extraordinary myth, with its story line that seems so brutal and naïve, in fact contains the most profound knowledge of the processes of the stars. It was from this knowledge that fundamental notions of the meaning of existence first took their shape. To understand this shape and how it is embodied in the myth, we must examine more closely the astronomical processes themselves.

The first organization of the world had been derived from the stable relationship that appeared to exist between the cycles of the planets, particularly the sun and moon, the unchanging patterns of the stars, and the determining role of the Milky Way. The myth of their separation, which put an end to this stable and golden age, tells, in a bewildering variety of images, of the awful realization that this harmonious relationship was not a permanent one. Due to the obliquity of the earth's axis, described in the previous chapter, over long periods the bonds that tied earth and sky together could be seen to become stretched beyond endurance. The gods themselves were being removed from their positions of authority, and the created world must eventually fall prey to the monsters of chaos and shake itself to pieces.

The fundamental consequence of the obliquity of heaven and earth is that the planet Earth behaves rather like a child's spinning top. Because the earth's axis is tilted, its extremities, which we see marked by the pole-stars and which we call the celestial poles, north and south, describe, over periods of thousands of years, a circle around a central point. This central and unmoving point marks what Milton called "the sun's axle," the axis of the wheel of heaven around which the planets, including the sun, travel, along the path we call the ecliptic. The poles of this central unmoving axis are known as the northern and southern ecliptic poles. As a result of the movement of the earth's axis of rotation around this central point, any star selected to mark the celestial pole cannot do so forever. Gradually the pole shifts, along the circumference of a 23.5-degree circle around the ecliptic pole, until it comes closer to another star than to the first.

What is more, it is not only the polestar that is displaced. The stars that formerly marked the solar stations at the corners of the square earth, the equinoxes and solstices, are bound into the Frame of Time by the colures. Since this construction is a rigid one, all its elements, each firmly linked in its relationship with all the others, must move as one.

It follows that when the polestar falls as the polar axis shifts, so must the stars at the corners of the earth also be demoted and new ones elevated to their glory. In fact, solstices exist only as a result of the obliquity of the polar axis and the consequent separation of the planes of the celestial equator and of the ecliptic. They represent the arrival of the sun at that position on its path that is farthest from the celestial equator. The equinoxes occur when the sun crosses the celestial equator on its way north toward the summer solstice and again on its way south toward the winter solstice. Were the axis of the planet earth vertical to the plane of its orbit, then would "spring perpetual" reign, to use Milton's phrase, and the sun's path would be along the celestial equator.

This is an important point to appreciate, since it rules out the possibility that the myths describe the "historical" event of the displacement of the earth's axis from the vertical. The myths are not records of cosmic catastrophes or collisions. If the world was created with four corners, then it was created with the obliquity in place. We shall soon be in a position to appreciate the true significance of the mythical "separation."

As we have seen, the polar axis is in fact at an angle of 23.5 degrees from vertical relative to the axis of its orbit, and as a result, its extremities move around the unmoving pole of the ecliptic. So too must the two points (the equinoxes) that mark the intersection of the equator (the earth) with the ecliptic (heaven) move around the circle of the ecliptic, passing through each of the zodiac constellations that stand along it, one by one.

This is the phenomenon known as the precession of the equinoxes. (The direction of movement of the intersection relative to the fixed stars is the opposite of that of the annual journey of the sun.) It is perceptible over long periods by observing the stars that rise just before the sun at the equinoxes. A change of one degree along the circle of the zodiac occurs (according to the traditional, though approximate, measurements) every 72 years, meaning that to pass through an entire zodiac sign of 30 degrees will take approximately 2,160 years.

For astronomers today, this phenomenon is a commonplace among the vastness of intergalactic space. For the ancients, it represented the fundamental principle of the law of existence. The cycle of the precessional ages is what the Hindu priests knew as the "third mystic wheel of the sun," the Wheel, as Stutely and Stutely call it in their Hindu dictionary (although they do not identify it as related to the precession), of the transcendental world, which they say is described in the Rig Veda hymn as "known only to those skilled in the highest truths."[1]

To form an image of this immensely slow process is for us enormously difficult. It takes prolonged thinking about the sky in terms of this kind of model before the elements begin to form themselves into an intelligible order, until the theoretical model of the diagrams can be discerned in the actual movements of the stars. It is clear, however, that from the most ancient times, such exposure was the norm at least for certain selected members of the population, if not for everyone. The almost

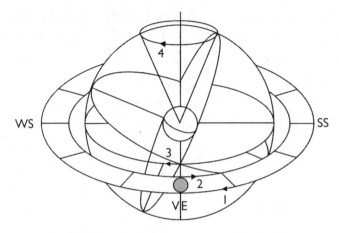

1. the daily wheel of the sun across the sky
2. the yearly wheel of the sun through the Zodiac
3. the "mystic wheel of the sun," the precession of the
   equinoctial sun through the zodiac
4. the circle described by the celestial pole during
   the precession

*Fig. 14. The Wheel of Transcendent Reality. In this diagram, the horizontal
circular plane represents the plane of the Earth's orbit around the sun,
known as the ecliptic. The vertical axis is the ecliptic pole.
"Your two wheels, Surya, the Brahmins know in their measured rounds. But
the one wheel that is hidden, only the inspired know that."*
—RIG VEDA HYMN 10.85–16

unimaginable amount of effort that went into erecting the earliest build-
ings, all of which embody the knowledge gleaned, together with the vast
amount of decorative and, later, written records of celestial events, reveal
the obsessive manner in which this work was pursued.

Only numerous generations, with a system of passing on their accu-
mulated information, could achieve the kind of understanding necessary
to execute such work. It now seems possible that the myths themselves
were this system. It is certainly clear that by the time systems of written
records appeared, a profound understanding had been well established.
The night sky must have been the object of intense study throughout this
time. The motive force for this obsession was clearly a powerful one.

THE BEGINNING OF TIME

The cycle of the precession is an endless one, repeating itself every 26,000 years, more or less. (The traditional figure is 12 × 2160 = 25,920. In fact, the figure varies, as does the precise inclination of the earth's axis.) The myth of creation, however, talks invariably about a beginning. When Saturn separated his parents, Heaven and Earth, he set in motion the machinery of time, with all that this entails, as an Orphic fragment tells us: "Saturn has become the cause of the continuation of propagation and begetting . . . from which originates the division of beings."[2]

This myth describes the moment in the cycle of the precession when measuring might be considered to have begun—the point in the cycle, in fact, that marked the birthday of the world. At first sight, it might seem that the choice of this moment would be arbitrary, since, as Aristotle said of the perfect sphere, "no line could be drawn between part and part within the whole." Nevertheless, as we have seen, the opinion persists around the world that the beginning was a particular kind of moment. In its first age, the world had been perfect. The original creation had been without death or disease or unhappiness. During this era, the Ancestors had laid out the landscape and introduced the arts of organization. Gods and Men conversed together, and heaven and earth had been in perfect harmony.

It is this perfect world that is ritually re-created in the New Year ceremonies. Indeed, the whole year is ritually regarded as a small-scale version of the much vaster cycle of the precession of the equinoxes.* Hence, this cosmic cycle is subdivided by the arrival of the equinoctial crossing of ecliptic and equator at the beginning of each new zodiac constellation. It can thus be said to have twelve "months," each of which is described as a new world, or a new age of the world (what the Hebrew scriptures call "Aeons" or simply "Times"). The whole cycle is referred to as the Great Year. Can we now say when New Year's Day of this Great Year might have been, the birthday of the First World?

---

*In the Bundahishn (a Pahlavi creation text), creation is said to have taken a year, after which the creator rested for five days.

To qualify, it would have to be a stage on the cycle when the "elements of creation" stood in an arrangement that could be said to constitute the most perfect relationship between heaven and earth. It would have to be the most satisfying location of all for fastening the earth to the sky.

Up until now, we have concentrated our attention on the divisions of the starry circle created by the solar seasons. To answer the question of when creation began, we must return to the other method of dividing the heavens that we introduced in chapter 6. This is the galaxy, that broad band of stars that we call the Milky Way, which is a "side view" of the spiral arms of the galaxy in which our solar system rotates. As we saw, it does present, both practically and traditionally, the most apparent means of dividing and thus measuring the circle of the stars.

To establish a situation that would deserve the title of "Perfect World," we need to find a way of combining this River of Heaven with the Frame of Time as defined by the solar stations. That is to say, we need to look for a point on the endless cycle of the precession when one or the other of the solstices or equinoxes, the "corners of the earth," occurred at the points on the sun's path where it crosses the Milky Way.

These two points, between the stars of Gemini and Taurus at one "end" of the Milky Way and between Sagittarius and Scorpio at the other, were first proposed as the foundations of creation by Santillana and von Dechend. They suggested a date of 4300 BCE for the "beginning of time," since at that time the spring equinox sun stood at one of these points (between Gemini and Taurus), and this date coincided with the rise of the Mesopotamian traditions, within which they considered the myths to have arisen.

Since that time, Bauval and Hancock, investigating Egyptian texts, have proposed an earlier date, around 10,500 BCE, when the solstices occurred at these two intersections of ecliptic and galaxy (although Hancock and Bauval have other reasons for choosing this date).[3]

In establishing their date for the beginning of time, Santillana and von Dechend described the criteria by which we might identify it. "The

Fig. 15. Midnight, spring equinox, 4320 BCE. The Age of Taurus. The sun is in the constellation Taurus (at the opposite side of the sky). This image and those of figs. 17–19 are all as observed from the equator looking south; the equinoctial crossing is thus directly overhead (at "zenith").

Galaxy was and remains, the belt connecting North and South," they explained:* "But in the Golden Age, when the vernal equinox was in Gemini, the Milky Way had represented a visible equinoctial colure; rather a blurred one, to be true, but the celestial North and South were connected by this uninterrupted arch, the galactic avenue, embracing the 'three worlds' of the gods, the living and the dead."[4]

Just how blurred a "colure" this was can be seen in the diagram of the sky at this time (fig. 15). In fact, when the spring equinox was in or near the stars of Gemini, the Milky Way was as far from being a

---

*In the Vogul myths of Siberia, one of the two collaborating creators is said to have strengthened the unsteady earth by passing around it a silver belt with silver buttons.

north–south line as it ever gets, as the diagrams show. It is never less than 60 degrees or so from the pole and never reaches farther north on the horizon than northwest (fig. 16).

If we want to have the best possible location for the fastening of the sky, the emergence of creation, then we must turn back the precessional clock another 180 degrees so that the vernal equinox occurs in the Milky Way between Sagittarius and Scorpio, on the other side of the sky. At this point we will find a considerably more "perfect" moment to choose for the beginning of the world. It incorporates a phenomenon that occurs only once in any full precessional cycle of 26,000 years, which affords the best possible arrangement of heaven and earth. The date is

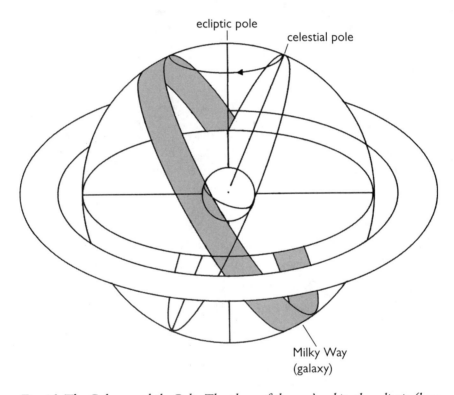

Fig. 16. The Galaxy and the Pole. The plane of the sun's orbit, the ecliptic (here shown horizontal), is at an angle to the axis of the galaxy, which is approximately the same as that of the obliquity of the earth's axis. The diagram shows the relationship between the celestial pole and the Milky Way in 4320 BCE. At this point, the celestial pole is as far from the galaxy as it ever gets.

Fig. 17. Midnight in the Garden of Eden. The midnight sky, spring equinox, 17,280 BCE; the Age of Scorpio. The sun is in Scorpio, at the opposite side of the sky.

around 17,280 BCE. During that era, for a period of perhaps a thousand years, the celestial pole was in the Milky Way.*

As a result, in such an arrangement, the Milky Way truly does become an avenue between north and south. The Great Circle joining the equinoxes with each other and with the poles follows the best possible line through this broad belt of stars. The two opposite corners of the earth could be seen "suspended" from the pole by this starry "bond of heaven."

The diagram shows how this perfect sky would have looked (fig. 17). Since the pole around which the sky turns is in the Milky Way, not

---

*This date is a "nominal" one. When first running back the software to establish the best possible position of the celestial pole in the galaxy, I reached 17,300 BCE. The traditional length of the precessional cycle is 25,920 years. If we take the moment between 1 BCE and 1 CE as our datum, precessing back eight "Great months" (8 × 2,160) gives 17,280 BCE.

only would this band of stars always be visible in the night sky, but it also would act rather like the hands of a clock, from which could be read the time of night and of year. It would be a veritable Frame of Time.

THE LADDER OF THE GODS

At the equinoxes during the First Age of the world, when the celestial pole was in this ideal position, the Milky Way reached the central point in its nightly journey across the sky, its culmination, at midnight. At that moment it would flow along a north–south line from horizon to horizon. This Celestial River has been recognized in many places as the heavenly echo of the local terrestrial river. The most famous example must be the Egyptian Nile, which was seen as a direct continuation from earth to heaven in the north and back again from heaven to earth in the south. This image must have been particularly potent along that stretch of the Nile between Aswan and Luxor, where the terrestrial river flows due south–north, just like its equinoctial midnight counterpart above in 17,280 BCE.

In this way, the Milky Way became regarded as a direct link between the earth and the very center of the heavens. But this link was imagined not only as a river. The Maya called it Sack Be, the White Road, and Xibal Be, the Road of Awe. Like many others before and after them (Chaucer calls it "Watling Street"), they regarded it as the road along which the souls of the dead returned to their heavenly home, as we shall see in the last part of this book.

Nowhere could the attritional impact of the precession be more noticeable than in a situation where a major river such as the Nile could act as the terrestrial counterpart of the heavenly river from horizon to horizon. Gradually, celestial and terrestrial rivers would part company as the pole moved out of the galaxy, until no passage from one to the other was possible, at which point earth would clearly be seen to be cut off from the route to heaven.

Sometime after I had identified this moment in the precession as the most likely candidate for the title of birthday of the world, I noticed a most surprising feature of the night sky of that era. Around the area of the Milky Way, where it crosses the ecliptic between Gemini and Taurus, are many of the most familiar stars in our sky. Four of these, Sirius, Pollux (in Gemini), Capella (in Auriga), and Aldebaran (in Taurus/ Hyades), form a majestic cross lying almost vertically along the Milky Way. The center of this cross lies exactly on the ecliptic, at the center of the River of Heaven. At the moment of creation identified here, this point lay also on the intersection of equator and ecliptic, the equinoctial point at the very heart of creation. This cross stood vertically in the sky at the spring equinox midnight during the First Age of the world. The Maya called it the Wakah Chan, the Raised-up-Sky, also referred to as the World Tree. They pictured it as a richly decorated cross, with the double-headed snake of the ecliptic draped across the arms.[5]

Although she does not acknowledge the precise location of this cross, regarding it as a general symbol for the Milky Way, Linda Schele explains that the Maya "thought of the entire north direction as a house erected at Creation with the World Tree, the Wakah Chan, penetrating its central axis."[6]

Only when the heavens are in the perfect position we have outlined does this really describe the situation visible in the sky itself. The conquering Spaniards recognized the decorated crosses that represented the Wakah Chan in the Mayan celebrations as examples of the foliated crosses common in European churches at the time. The Maya, for their part, readily accepted the Christian cross as a symbol of the central pivot of their world.

Santillana and von Dechend mention the medieval tradition that the wood from this central tree, the Tree of Life in the Garden of Eden, will one day become the material for Christ's cross. They go on to quote the Finnish runes, which describe the Great Oak: "Long oak, broad oak. . . . The sky is the root of its summit. An enclosure within the sky. A wether in the enclosure. A granary on the horn of the wether."[7]

Before we dismiss this wether in its enclosure as simple fancy, it would be good to remember that at the summit of the starry cross we have identified is Capella, the goat-star, also known as "the horn of plenty."[8]

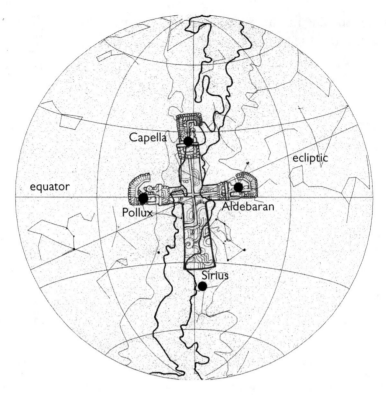

Fig. 18. The Raised-up-Sky. The midnight sky at spring equinox, 17,280 BCE, with the Mayan Wakah Chan superimposed. Mayan wooden crosses are frequently decorated with flowers at the positions of the four stars.

As an image of the direct route from earth to the heavenly city, the Milky Way then can be River, Tree, and Road. It can also be a ladder, a stairway to heaven. The Egyptian funerary texts make frequent reference to this ladder as the means of ascent to the stars: "Now let the ladder of the god be given to me, that I may ascend on it to the sky,"[9] says one of the ancient pyramid texts; and in the Book of the Dead, the triumphant supplicant exclaims, "I go onwards to the domain of the starry gods. . . . I set up a ladder to heaven among the gods" (chapter 149).[10]

The access route to the heavens can even be a rope, as in this account quoted by the classical Mayan scholar Alfred Tozzer: "There was a road suspended in the sky [it stretched from Tulum to Chichén Itzá]. . . . It was in the nature of a large rope, supposed to be living and in the middle flowed blood. It was by this rope that the food was sent to the ancient rulers who lived in the structures now in ruins [i.e., Chichén Itzá]."[11]

And this rope appears again in a story from central Celebes: "In the beginning the sky was very close to the earth, and the Creator, who lived in it, used to let down his good gifts to men at the end of a rope."[12]

Whichever form it takes, this link is how contact between heaven and earth is maintained, not just for people to reach heaven but also, as this rope-road reveals, for the gods and their treasures to reach the earth.

The Milky Way can be all these things. But, according to the myths, it can also be a dried-up river, a burned-out road, a chopped-down tree, or a broken ladder. Even the rope cannot last forever. The report that describes the Mayan Rope-Road goes on to say, "For some reason this rope vanished forever. This first epoch was separated from the second by a flood."[13]

In South American traditions, the Milky Way is said to be the ashes of a great tree that once contained all forms of life. For the Carib Indians and others, the tree had been the source of cultivated food, but one of the creator twins decided to chop it down, despite being warned of the consequences. When it fell, water spilled from the trunk and flooded the earth. In the same way, the Finnish traditions that tell of the Milky Way as a huge tree, "the great oak," explain how this tree fell across the sky and obstructed the passage of sun, moon, and stars. It was eventually felled by a tiny creature who emerged from the sea or from under the earth.

Of the images of the Milky Way as a primal, nourishing tree, perhaps the most striking is that reported from the Guatemalan Maya who live by the shore of Lake Atitlan:

> Before there was a world, a solitary deified tree was at the centre of all there was. As the world's creation approached, this deity became pregnant with potential life; its branches grew one of all things in the form of fruit. Not only were there gross physical objects like rocks, maize and deer hanging from the branches, there were also such elements as types of lightning, and even individual segments of time.

Fig. 19. The Fallen Tree. The midnight sky, spring equinox, 15,120 BCE, the end of the Age of Scorpio.

Eventually this abundance became too much for the tree to support, and the fruit fell. Smashing open, the fruit scattered their seeds; and soon there were numerous seedlings at the foot of the old tree. The great tree provided shelter for the young "plants," nurturing them until finally it was crowded out by the new. Since then, this tree has existed as a stump at the centre of the world. This stump is what remains of the original "Father/Mother, the source and endpoint of life."[14]

If it were not for this image of the new trees that overcrowd the original deified, time-bearing tree (these young plants sound like the "overbearing" Children of Heaven), we might easily miss the significance of the seeds brought to earth in the story told by the Mohawk/Iroquois people of North America that tells how the Woman in the Sky dreamed that the great Sky Tree must be uprooted. The Skyland Chief, who some say was suffering the first illness, had the tree uprooted and was thereby cured. Below, the earth was only water. Sky Woman fell (many stories say the chief pushed her) through the hole, clutching the seeds of the tree. The Muskrat brought up earth from the bottom of the waters and Great Turtle supported it. The Woman dropped the seeds of the Sky Tree in her footprints.*

As in many of these stories, here it is the first era of creation that has come to an end as this tree falls. We learn from other versions, however, that at the same time the direct route to heaven has been closed, as in the story of the first king of Tibet. He is said to have descended to earth on a ladder "like a rainbow." He left the same way when his son was old enough to rule, but the ladder was eventually severed from the earth. From then on, the descendants of the king had to find a new way to return to heaven.[†]

The Australians also tell of the felling of a giant casuarina tree, "which was ready to provide a suitable ladder for men to climb to the sky. But the tree was cut down by certain mythical personages, and the bridge between heaven and earth was destroyed for ever."[‡]

In Australia, stories similar to this frequently tell that it was at this moment of separation that death entered the human world for the first time: "Man had to die only because all connections had been severed between the sky and the earth."[15]

The Achilpa people, one of the Aranda groups of central Australia, told this story slightly differently. During the first creation, Nambakulla ("Arose Out of Nothing") traveled north, making mountains, rivers, and "spirit children." He taught the ceremonies to the first ancestors he

---

*Various extended versions of this story are related in J. N. B. Hewitt, *Iroquoian Cosmology;* www.sacred-texts.com/nam/iro/irc/index.htm.

†This king was the first Tsampo (king), the founder of the Tsampo Dynasty, which ruled Tibet for centuries. It is tempting to relate this word to the Rig Veda's "Skambha" and the Finnish "Sampo."

‡Mircea Eliade, *Australian Religion* (Ithaca, N.Y.: Cornell University Press, 1973, 31). Eliade adds, "Strehlow reminds us that traditions about broken 'ladders' have been found at many ceremonial sites."

made: "Now Nambakulla had planted a pole called *kuwa-auwa* in the middle of the sacred ground. After anointing it with blood he began to climb. He told the first Achilpa Ancestor to follow him, but the blood made the pole too slippery, and the man slid down. Nambakulla went on alone, drew up the pole after him and was never seen again."*

These are just a few of the array of stories about the severing of ropes and bamboo stalks, of trees and ladders, of dried-up rivers and burned-out paths, that tell of the end of direct communication with the gods. The stories may differ, but the end result is the same. The first creation has come to an end engulfed in the devouring Flood. No longer will the galaxy act either as a nightly clock or as a stairway to heaven. "The links between Heaven and Earth were broken," as the myth says; death has arrived, and the world can never be the same again "until the Times Shall Be Fulfilled."

The Milky Way is a wide river. It could be regarded as occupying this crucial role in defining the heavens for more than a thousand years. It must have been a long process, then, the realization that it was not an eternal arrangement, but ultimately the awful truth would have been unavoidable. The polar axis, about which the sky nightly turned, was shifting. It was, perhaps, the gradual drift of the pivot of the sky away from the Celestial River that first revealed to patient but increasingly anxious observers that all was not well with the order of the heavens. At the same time, inevitably, the halves of the year, as divided by the equinoxes, began to part company with the halves of the sky, divided by the Milky Way.

Any society that had constructed its whole system of organization around the model of a celestial world that had endured more or less unchanging for a thousand years would be forced to acknowledge the ravages that time could wreak on their celestially ordered lives. "For our ancestors," says E. C. Krupp, "what went on in the sky was a meta-

---

*Mircea Eliade, *Australian Religion,* 50. Eliade relates that the Ancestors then traveled continuously, carrying out rituals. The *kuwa-auwa* was always erected and made to lean in the direction in which they intended to travel. A representation of this pole was erected during the long series of initiations of the Achilpa.

phor. It meant something. It was both the symbol of the principle that they felt ordered their lives and the force behind those principles. . . . They sensed that the sky orders our psyches and our societies, and they expressed the bond between brain and sky in their works."[16]

We can only speculate on the impact of this dawning of decay on those people who had come to identify the stability of their own sense of order with the apparent order of the heavens, now crumbling before their eyes. To me, it seems the most astonishing of all human achievements that, if such was their experience, they managed to construct out of it not a message of disaster and social disorder but an even more magnificent structure of order, order on a truly grand scale.

ONE AND ALL

THE BODY OF GOD

It is quite possible that stories of a First World were never more than nostalgia, that it was not until the effects of the precessional process had been discerned and described that it became clear that the elements of creation had once been in that perfect relationship. That is to say, astronomers at any date later than 17,280 BCE, equipped with the required understanding of the precession, could "wind back" the precessional clock until they reached a vision of that point in the past when the celestial pole stood in the Milky Way.*

However, the immense significance that every tradition has attached to the passing away of the perfect world, and the profound and lasting effect it has had on human thinking, strongly suggests that it was an observed event. Indeed, any society that derived its sense of order from a sky understood to be both balanced and stable would be ideally

---

*William Sullivan considers the adoption of the precessional mythology as concurrent with the arrival of agriculture, based on the notion that such an economy would demand a sophisticated solar calendar. This would put a limit of around 10,000 BCE on the date, according to current archaeology. Sullivan does suggest, however, that star-watching was a pursuit of hunting societies. More-recent discoveries within the caves of Lascaux have been claimed to reveal detailed knowledge of the important star groups by 14,500 BCE (see the work of Michael A. Rappenglück at www.infis.org).

placed to observe the havoc that the precession could work on its celestially ordered lives.

According to the myths, their first response was to establish a new "creation." If the Milky Way no longer determined the parameters of their organized world, then some replacement must be found, and the most immediate claimants were seen to have been Rigel (Orion's left foot) and the Pleiades in Taurus, stars that, together with their "partners" on the opposite side of the sky, Vega and the claws of Scorpio, marked the equinoctial colure in this new age. But the precession was a relentless mill, and it would inevitably grind down any established world. Even the traditions that appear to describe only one world existing after the destruction of the Flood, such as that contained in the Hebrew and Christian scriptures, nevertheless make reference to the idea of a "new heaven and a new earth," and talk about the "fullness of Time," that is, of the completion of the "Ages" (Greek Aionon).*

What the ancient observers began to conceive, beyond the destruction of their former world of perfection, beyond the inevitable fate of any subsequent creation they might identify, was the notion of a cycle of creations and destructions. And this cycle could be imagined to have its own existence, an existence unlike any other in the perceived world. A complete cycle would embrace all the potentialities simultaneously; it would exist outside of time, "changeless and unsullied," as the Hindu text puts it, free from the degradations to which the manifested worlds were prey.

We have a tendency today to speak about "eternity" without giving it much thought, using it vaguely to mean "forever and ever." The eternal, however, is not so easily grasped. The finest theological minds throughout history have struggled to encompass with words the supreme being and its indescribable nature. And yet it has inspired the most profound

---

*In translations of the Bible, this term "Ages" receives a number of interpretations: it can be "the times," "since the beginning," and even "everything there is" or "the whole of creation." Something similar is contained in the idea of the Catholic Missal's ending of prayers, *saeculum saeculorum*, translated as "for ever and ever" but meaning more literally "cycle of cycles."

of human achievements and aspirations. We might well ask how an idea so powerful and at the same time so difficult to comprehend first found expression in the human mind. In the discernment of the full cycle of the precession, I believe we can find an answer to this question.

What is certain is that since that moment when the image of the Great Year as the summation and surmounting of time was first promulgated, it has determined the form of all speculation concerning the nature of the world and our position in it. This is as true today as it was for Plato. In 1980, for instance, physicist David Bohm published a book called *Wholeness and the Implicate Order*. In it, he proposed an immeasurable, unknowable order that prevails in some multidimensional space and that manifests in various subtotalities. Each of these subtotalities is bound together by a "necessary force," but each must eventually dissolve back into the universal. Bohm may have been able to subject his Implicate Order to more-detailed mathematical analysis, but it remains nevertheless a close descendant of that eternal completeness described by Anaximander, a Greek philosopher of the Aristotelian school in the sixth century BCE: "The first principle of existing things is the Boundless for from this all things come into being and into it all perish. Wherefore innumerable worlds are brought to birth and again dissolved into that out of which they came."*

This description takes us back to that principle from which we saw creation first arise, the "sea of potentiality," within which "the Lord revealed himself, irresistible, self-existent, subtle, eternal, the source of all beings." We can now appreciate the meaning of that event which in the Hindu myth is described as the "awakening" of the divine energy. It represents the moment of recognition of the entire precessional cycle of "worlds." This subtle, eternal source of all beings, this boundless source of creations, we might truly call the Body of God.

---

*Arthur Fairbanks, ed. and trans., *Anaximander, Fragments and Commentary: The First Philosophers of Greece* (London: K. Paul, Trench, Trubner, 1898, 8–16) (at history.hanover.edu/texts/presoc/anaximan.html). Anaximander is said to have added that "Gods have a beginning, at long intervals rising and setting, and . . . they are the innumerable worlds."

We have already met the Egyptians' embodiment of this idea, the deity Atum, whose name, according to Wallis Budge, means "wholeness" or "completeness," "he who had come into existence as a circle." Valentinus, a Christian Gnostic of the second century, who held some unorthodox views, referred to this "Dweller in the Primeval Waters" as the "true divine source," calling him the "Depths" or the Bythos (Abyss). But this is not the abyss of chaos, where the opposites rove unfettered and disorderly motion reigns. This is the fullness of potential, but arranged in its circular order, according to the "Path of the Wisdom of the Highest," as Plato puts it in the passage where he describes this vision, leaving us in no doubt as to its nature: "The form of the godhead He consecrated and made for the most part of fire, that it might be the brightest of all and fairest to look upon; and likening it unto the Universe He made it spherical, and set it in the Path of the Wisdom of the Highest, to go therewith and distributed it over all the spangled round of Heaven, to be a true adornment thereof."[1]

This, then, is the ineffable, incomprehensible Godhead, who is beginning and end, who was, is, and ever more shall be. Or rather, as Parmenides puts it, "Being is without beginning and indestructible; it is universal, existing alone, immovable and without end; nor ever was it nor will it be, since it now is, all together, one, and continuous."[2]

Out of this primal source come the created worlds, each of which will ultimately be dissolved back into the ocean of potentiality. All the stories we have of "gods" and "creations" are but endless elaborations and detailed examinations of this process, as Plutarch says:

> God is of his own nature incorruptible and eternal, but yet through a fatal decree . . . suffers changes . . . having sometimes his nature kindled into fire, and making all things alike, and otherwise becoming various . . . like unto the World and is named by this best-known of names. But the wiser, concealing from the vulgar the change into fire call him Apollo. . . . As for the passion and change, this they obscurely represent as a certain distraction and dismembering and they now call him Dionysus, chanting forth

*Fig. 20. The Boar Raises the Universe from the Waters. Detail from
Atha Srimadvarahamahapuranam (Varha's Great Ancient Tale),
Kalyananagaryam: Laksmivenkatesvara Mudranalaye, 1923.*

corruptions, disparitions, deaths and resurrections which are all
riddles and fables.*

And one of these "fables," retold in the Hindu Vishnu Purana
hymn, is a clear description of what it is that drives the transformations
that Plutarch describes:

Realizing that the earth was within the waters when the universe had
been made into a single ocean, Prajapati [a form of Vishnu known as
the "Man," from whose body the universe had been made] wished to
raise it. He made another body; as at the beginning of previous aeons
he had made the fish, the tortoise and others so now the eternal, constant, supreme soul, the soul of the universe, Prajapati took the body

---

*Plutarch, "Of the Word EI Engraven Over Apollo's Temple at Delphi," *Plutarch's Morals*
vol. 4 (available at oll.liberty,org). Guthrie (*History of Greek Philosophy*) gives a summary
in which he replaces "riddles and fables" with "mythological tales." The details of the text
are susceptible to interpretation, but the general sentiment is clear.

of a boar, a form composed of the Vedic sacrifice, in order to preserve the whole universe.³

We learn not only that it is creation itself that is repeated but also that it is a succession according to aeons; it is time that dictates the procedure. We have already learned what Vishnu told Arjuna: "Know that I am Time, that makes the worlds to perish when ripe and brings on them destruction."

THE MOVING IMAGE

And this is how it must be for creation, which can exist only in time, separated from the undivided eternity. The waters, as Eliade described them, "can never pass beyond the conditions of the potential of seeds and hidden powers. Everything that has form is manifested above the waters, is separate from them. . . . As soon as it has separated from water . . . every form . . . falls under the law of time."⁴

Regardless of the intentions of a Demiurge, therefore, creation is a frustrating business. Valentinus describes how the creator wanted to make manifest the perfection of the plan that he could envision by means of the power of reason. In this plan, all the elements, arranged in harmonious pairs, coexisted in equilibrium. The Demiurge, says Valentinus, "desiring to imitate freedom from all measurement, being unable to express eternity, had recourse to the expedient of spreading out eternity into times, seasons and vast numbers of years."⁵

What is more, this kind of creation was flawed and was thus doomed to pass away, as Valentinus goes on to explain: "Truth having escaped him . . . he followed what was false and when the times are fulfilled his work shall perish."

To the Christian Church, as it began to formulate itself from the fourth century onward, this was (and still is) a horrendous idea, that God was somehow imperfect, at the mercy of his materials, so to speak. St. Paul also struggled with this notion, although he concluded from the text of Genesis that this condition was the result of "the Fall" and thus foresaw a time beyond: "For the creation was subject to futility, not out of its own will but by the will of him who subjected it in hope, because the creation itself will be set free from its bondage to decay and obtain the glorious liberty of the Children of God."[6]

The origin of this problem is deeply ingrained in the texts of the Hebrew scriptures, which have come to confuse the all-embracing One, the Unity, with that One which is the first of many. As Marie-Louise von Franz points out, "This paradox is already contained within the Pythagorean cosmogony in the idea that the one, as the monad, sometimes represents the one original 'arche' of the world and sometimes reveals itself as the generating seed. Instead of assuming a division of the monad, I prefer to consider the latter running right through the whole number series. One can compare it to a 'field' in which the individual numbers represent activated points."[7]

In Valentinus's opinion, the creator God described in Genesis is but an anthropomorphic image of the "true divine Source" underlying all being, which Valentinus called "the Depths." This image he refers to as king and "Demiurge," regarding him as a kind of craftsman, fashioning the universe as best he could. He gives him the name Yaldabaoth, "the serpent-lion," and explains how he had "caused destiny to come into being, and thus bound the gods of the heavens, the angels, daemons and men by means of measure, epochs and times, so that all came within destiny's fetter."

In Greek tradition, the origination of destiny is ascribed to Kronos, as a result of his "crime," the separation of his parents. In Hesiod's view, by this act "the obstacles to genesis are removed, space is opened out and the world becomes organized . . . but at the same time the *dolos* [trick] is a terrible lapse, a crime committed against the primordial powers [Heaven and Earth while they still remain unseparated] which are at the origin and source of all existence."[8]

Valentinus seems to have received his ideas on this subject from Plato, although Plato does not go so far as to call the act of creation a crime. Rather, he explains that the Demiurge was a victim of his circumstances. Plato describes how, in fashioning the world, God had worked with preexistent materials that he was powerless to alter. These materials suffered from *ananke*—"brute necessity"—the characteristic that distinguishes "what ought to be" from "what cannot be got away from." It is the craftsman-god's achievement to fashion an ordered universe from brute matter despite its inappropriateness for the task. Plato explains how this was done: "Because the pattern of the Universe was an Eternal Being, whose nature was eternal, and which could not be joined to a created thing, He took thought to make a Moving Image of Eternity . . . progressing according to number, to wit, that which we have called by the name Time."[9]

THE FOUR AGES OF THE WORLD

"By definition," says Marie-Louise von Franz, "the great cosmic unity cannot be subdivided into a plurality of 'quantities'; it can only reveal itself in the course of time (always as the same whole) in various qualitative aspects . . . which are circumstantial—literally translated as that which stands around."[10]

Each creation that comes from the eternal is a "qualitative aspect" of the supreme god. It is ruled by the stars that determine its measure, the "circumstances," we might say. Put simply, they are the polestars and the four stars (or groups of stars) that stand at the four corners of the square earth and mark the solstices and equinoxes. Of these six, two are commonly given the greatest significance. They are the North Pole star and the group of stars that stands at the beginning of the year at the crossing of the equator and the ecliptic and marks the spring equinox.

Either of these two (or sometimes both) may appear in the myth as the "king." In addition, those stars that mark the year's beginning may be referred to as "leader" or "foundation." Such titles may thus be applied to a number of different star groups, as each one reaches the "position of authority." The Assyrian word *ku* means both "prince" and "foundation," and was applied in successive eras to the stars of Auriga, of Aldebaran in Taurus, of the Pleiades, and of Aries as they took up their positions as seat of the vernal equinox.

We have seen how these frame-defining stars, bound together by the Great Circles of the equinoctial and solstitial colures, constitute what the myths call "the world." It is within this world that earthly societies exist. Its measures define the ritual calendar, the social order, and the political ethos of the age. Any calendar based on the "perfect" arrangement, starting the year when the sun reached the galaxy, would slowly become disassociated from the solar seasons. The equinox would occur before the sun reached the galaxy, and New Year's Day would come increasingly late as the age wore on. This is the significance of the "breakdown of order." The ruling stars will no longer rise at the appropriate times, sacrifices will be held in a "disorderly fashion," and the law of Rita, the cosmic rule, will be ignored.

To put matters right, the world must be made anew. If creation is not to disintegrate into formless chaos, a new image of eternity must emerge from the boundless. As the crossing point of equator and ecliptic "precesses" from one zodiac constellation into the next, the old frame must be dismantled and "dissolved into the boundless."

The manner in which the stars exercise their authority is a complex one, but we can see it operating at its most basic, for instance, in the age when the stars of Taurus marked the spring equinox. At that time the bull was the determining image for state and ritual alike. This was the biblical age of Abraham and the veneration of the golden calf.

As the machinery of time ground on, these images were replaced with those of the ram of Aries and later with the fish of Pisces. There are twelve "worlds" in the complete cycle, one for each zodiac constellation. The actual size of the constellations varies, but the whole circle of 360 degrees is divided into twelve 30-degree segments. For the equinoctial point to pass through each sign takes approximately 2,160 years. These "months" of the Great Year are traditionally grouped into four "seasons," periods of 6,480 years made up of sets of three "months."*

---

*Some traditions, notably the Mayan, describe five "suns," one for each of five creations. The Mayan system was a particularly complex one, involving cycles of the planet Venus as well as Saturn and Jupiter and a counting system based on cycles of thirteen. This is the notorious cycle due to reach its culmination with the December solstice in 2012.

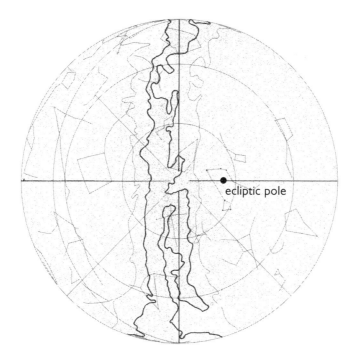

Fig. 21. The Celestial Pole; midnight, spring equinox, 17,280 BCE (and 8640 CE). South is at bottom.

These seasons are determined by the position of the Milky Way. The first starts with the spring equinox in Scorpio, as we have seen. The second is marked by the arrival of the summer solstice at this point, the third by the autumn equinox, and the fourth by the winter solstice. It is this final arrangement, the beginning of the last stage of the cycle of creations, the winter season, that we stand on the verge of. The best traditional date for the beginning of this age, when the spring equinox arrives at the constellation of Aquarius, is 2160 CE. Although the solstices began to move into the Milky Way around four hundred years ago (depending on how you measure the breadth of the galaxy), 2160 CE will see them arrive at the central point.*

---

*Again, this is a "nominal" date, counting "Great Months" of 2,160 years from the datum of 1 BCE/1 CE. Because the cycle is actually nearer 25,770 years, this date for the new age is slightly late.

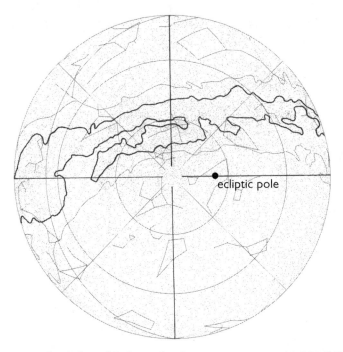

Fig. 22. The Celestial Pole; midnight, spring equinox, 10,800 BCE.

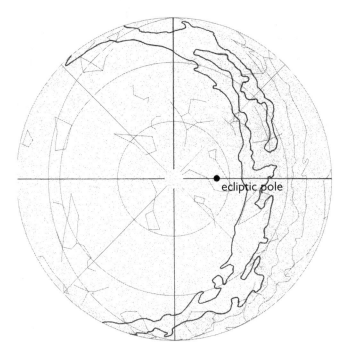

Fig. 23. The Celestial Pole; midnight, spring equinox, 4320 BCE.

Fig. 24. The Celestial Pole; midnight, spring equinox, 2160 CE.

In Christian tradition, each of these four epochs is introduced by one of the four evangelists, each of whose images is derived from the stars marking the spring equinox at his epoch's beginning. The Eagle, the stars of Aquila (or perhaps Cygnus), stands in for Scorpio (traditionally associated with death, for reasons that will become apparent, and thus inappropriate for this purpose) and the first epoch, when the spring equinox first began to leave the Milky Way. The Lion sees the equinox arrive at the stars of Leo, and the Bull at those of Taurus. In our era the Winged Man (the constellation of Aquarius, the Water Bearer) ushers in the last epoch of the cycle as the spring equinox arrives at the stars of Aquarius. Again according to Christian tradition, it is at the end of this epoch that "the times shall be fulfilled" and the last trump will announce the day of judgment.

BEING AND BECOMING

THE REAL WORLD

In acknowledging the passing of their "perfect" world, the ancient observers also recognized that the order of the world itself was mortal. The sky, once seen to be laying down impartial laws uninfluenced by the petty trials of people, had, in fact, a great bond with the people. They shared the same fate, to fall prey to the ravages of time. It was perhaps the recognition of this shared fate that gave the developing imagery its power and led to the obsessive compulsion with which it was pursued, embellished, and refined across the generations. This bond was to be of the greatest significance to those searching for a new place within the changing order around them.

The greatest of these observers were able to conceive of an eternal stability beyond their changing world. They were able to describe a greater pattern that could make something more of existence than a meaningless succession of change and decay. And if the created universe not only shared their mortality but could be seen also to exist outside the forms of division, in a timeless continuum, then surely people themselves might have a share in that eternal world.

From this insight, the human imagination was to take what must be its greatest leap. Out of what seemed to be the devastating understanding of the impermanence of the heavens was constructed an idea that lies at the very heart of whatever may be thought about the meaning of our lives. So fundamental to our way of thinking is it that we rarely give a thought to its origins, unless it is to suggest that it must be somehow built into the structure of mind itself.

Not just throughout history, but throughout all the peoples of the world as well, the idea persists, often in the face of logical argument on the one hand and fierce persecution on the other. This world is not our final home but rather a passing thing. Those in it are only temporarily so; their ultimate destiny is to find their way back to the Real World of the timeless whole. In Borneo, the Dayak Indians know that "the real native village of mankind is not in this world. Man dwells only for a time in this world, which is 'lent' to him."[1]

For the Dayak, the "real world" is the sacred land, and they can speak of it in some detail. It is where the sacred people live: "It lies among the primeval waters, between Upperworld and Underworld, and rests on the back of the watersnake."

Now, this sacred land, in fact, is none other than that described in different, more abstract terms by Plato, who calls it the world of Being, where the Ideal Forms eternally coexist. This is the world that contains "what ought to be," those things "made by the craftsmanship of reason." It is apprehensible by reason and thought, says Plato, since it is "ever uniformly existent." The "visible" universe, on the other hand, is in a state of "Becoming" and contains the things that "cannot be got away from." It is apprehensible by sensation. The things it contains are only "corrupted and corroded" copies of the Ideal Forms; "it comes into existence and is generated and perishes and is never really existent."[2]

Plato gives a description of the relationship between these two worlds, the real and the temporary, in the passage in which he tells of the manner in which God fashioned the Soul of the Universe. Because this passage is widely recognized as being fundamental to the development of Western metaphysical thought, it is worth considering in detail. Plato explains that the creator, "finding the whole visible sphere not at rest but moving in an irregular and disorderly fashion, out of disorder he brought order."

In the *Timaeus,* this disorderly motion is said to be the result of "opposites neither alike nor evenly balanced." He calls this state *alogos*— "without reason (i.e., ratio) or proportion." There had been in existence, he says, "fire and water, earth and air, showing traces of themselves but in such a condition as might be expected of anything from which God is absent."[3]

Plato then explains how *logos* was introduced into this disorder and "proportion" achieved: "Betwixt that substance which is undivided and always the same and that which cometh into being and is divided into bodies, he made by mixing them a third substance in the middle between the Same and the Other.* These substances, being three, he took and mixed all together, so that they became one Form, and the Nature of the Other, which was hard to mix, he joined by force unto the Same, and these he mingled with the Third Substance, and of the three made one."[4]

These portions, all standing in specified numerical relation to one another, are cut off in order until the whole "soul-mass" is used up. They are then pieced together in the order in which they are cut off and make a "soul-strip," which is then divided lengthwise. The two parts are then laid one across the other in the form of an oblique cross, the point where they cross being the middle of each. Each of these bands is then bent into a hoop so that its ends meet at the point that is opposite that at which the bands cross each other. One of these hoops is the Circle of the Same, the other is the Circle of the Other. The former revolves from left to right (east to west, the direction of the daily rotation of the stars and, incidentally, of the precession), the latter from right to left (west to east, the direction of apparent travel of the planets through the stars).

Plato then follows with a description of the numerical ratios that determine the proportions of this universe, the details of which need not concern us. The basic elements are clear: the Same and the Other, or "Different," as it is sometimes translated, when joined and laid across each other in an "oblique cross," are formed into the circles of the equator and

---

*We shall return to this mysterious third substance. It turns out to be nothing less than the "Mother of All."

the ecliptic. The whole Soul of the Universe is the model of the cosmos as we have so far constructed it. The fact that it took "force" to achieve the mixing of the Same and the Other was decisive. Their uniting is the description of the "incarnation" of the universe. It is a fall from grace: "Having before him physical material [the "unwrought matter of Time" is what Plato calls it elsewhere] he orders it, order being better than disorder, and thus finds on his hands something that can be called a body."[5]

In the *Statesman,* Plato explains further: "What we call universe and cosmos has received many blessed gifts from its creator, but nevertheless it partakes of body and so cannot remain for ever without change."[6]

It may help clarify this notion if we take this "body" (the word is *soma,* the Greek word the apostle Paul uses in the Christian scriptures when he asks whether the resurrection will be in the body or the spirit) of which the cosmos partakes to be the structure that we have elsewhere called the Frame of Time. Only one "element" of the eternal can be forced into this frame at any one time. Indeed, in other texts, Plato suggests that it is this forced conjunction that dislodges the harmony of the Soul. The "embodied" cosmos thus can manifest only one aspect of the Soul, the remainder of which rests in the eternal. With the demise of its mortal body, this aspect of the eternal will return to its source in the Real World.

THE VEIL OF ILLUSION

The majestic model of Being and Becoming that Plato offers (and we have only scratched the surface of it here) stands at the apex of the tradition he inherited, a tradition that he suggested came to him from Egypt. It has been the source of innumerable variations; the relationship between the One and the Many has entranced philosophers since time began, it seems. The Brahmavaivarta Purana calls it "the natural twofold division—that which is constant and that which is momentary."

In the East, this twofold division became the basis of the philosophy centered on the notion of *maya* (illusion). The word has the same root as "measure" (*matra*) and is applied to the perceived world. Thus, "the entire structure and order of forms, proportions and 'ratios' that present themselves to ordinary perception and reason are regarded as a sort of veil covering the true reality which cannot be perceived by the senses and of which nothing can be thought or said."[7]

In boldly attempting to say something of this true reality, David Bohm points to the fact that

> [the] notion of "measure," as the Greeks imagined it, is an inner quality of things, rather than our comparison with an external unit. It is an insight created by the forming activity of mind. To suppose that measure exists prior to the application of this activity leads to the "objectification of insight." . . . The men who were wise enough to see that the immeasurable is the primary reality were also wise enough to see that measure is an insight into a secondary and dependent but nonetheless necessary aspect of reality.[8]

To identify this "secondary reality" with the primary, immeasurable reality, however, "this is illusion." Meditation, we are told, is the process whereby this illusion can be repaired. The word *meditation,* whose Latin origin means "to cure" (cf. *medicine*), ultimately derives from the same root as that of *measure.* The object of both medicine and meditation is to restore the balance of inner measure to something that has "gone beyond its proper measure" and is therefore "out of harmony."

The source of all these formulations, and the means whereby this "healing" process is achieved, is revealed in the imagery and employment of the Chinese divination board. This ancient device consists of two boards, one circular, the other square, linked by a central pin, with the round board on top. The square board, known as the "Earth," is inscribed with the eight trigrams in their "time-bound" arrangement. This arrangement is known as the "Inner World" or "Later Heaven." The circular board is the Board of Heaven, said to be a "timeless structure." It is inscribed with the trigrams in the arrangement known as the "pre-World" or "Earlier Heaven." The interplay of the two boards signifies the "interplay of heaven and earth."

"On the *kanyu* the male is slowly moved in order to know the female."*

From its context in the Huainanzi, it is clear that some type of astronomical instrument is being manipulated. It may be the *shipan,* or cosmograph—an ancient planisphere—that is indicated here, models of which have been discovered in Han dynasty tombs. The specific meaning of *kan* is "canopy" and that of *yu* is "chassis"—reminiscent of the Chinese chariot—in which case, the term would mean "heaven and earth" or the "cosmos."

Around both the square and the round plate are arranged the names of the twenty-eight constellations of the Chinese zodiac. The twelve months are arranged by decimal number counterclockwise inside the ring of the heaven disk, whereas the twelve Earthly Branches, representing double hours of the twenty-four-hour day, occur clockwise on the inner square of the earth plate. In the center of the disk is a representation of Bei Dou, the Northern Dipper (known in the West as the Big Dipper or Ursa Major, the Great Bear). The dial of the cosmograph as it is rotated clockwise on the earth plate corresponds to the arc made by the stars as they pass toward their setting in the west. By means of the cosmograph, the configuration of the heavens could be determined at any time of day or night for any month during the year.

The imagery of this divination device is revealing enough. A description of the use of such a device tells how "the operator stood up, adjusting the board with his hands. Raising his eyes to heaven he gazed at the light of the moon. He looked to see where the Dipper was pointing . . . where the sun was situated. As auxiliary aids he used a pair of compasses and a set-square, together with weights and scales. Once the four nodal points were fixed and the eight trigrams were facing one another, he looked for the signs of good or evil fortune."[9]

This extraordinary process, which sounds like a precise reenactment of the creation, was considered to establish a link between "the One and

---

*From the astronomy chapter of the former Han dynasty Taoist text Huainanzi, 206 BCE–25 CE. Quoted in Marie-Louise von Franz, *Number and Time,* 235 ff.

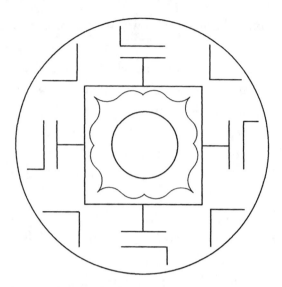

Fig. 25A. A schematic layout of a shipan, or "cosmograph," a Chinese divination board, Han dynasty (ca. 200 BCE)

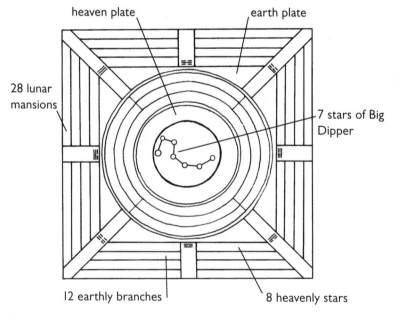

Fig. 25B. Diagram of layout of "TVL" mirror. "The T's comprise a vertical prop which supports a horizontal beam; the four V's mark the four corners of heaven. The four L's symbolize the ends of the two cosmic lines." (Michael Loewe, Ways to Paradise*)*

the Many, the coming together of an eternal order with the just-so-ness of reality, the sacred marriage of Heaven and Earth." It is just such an arrangement that was made permanent in the bronze mirrors that were described in the section "Surveying the World."*

THE SECRET OF BECOMING

We are thus at the very heart of how the wisest people of the ancient world constructed their idea of the notion of meaning. Looking at the night sky, they perceived the long, slow announcement of the laws of existence. This message they decoded by observing the "isomorphisms," or correspondences of form, that existed between what they observed in the sky and what they observed around them and felt in their hearts. In this way meaning was constructed.

"The secret of Being lay displayed before their eyes," as Santillana says.[10] And as the heavenly display unfurled, revealing its secret, so apparently devastating for the concept of order and stability, these "exceptional men" began to understand also the secret of Becoming.

And so began the development of the idea that has driven philosophers ever since. First they had perceived that the motions of the heavens were a declaration of the laws that governed every aspect of their existence. Then they discovered the mightiest of these laws, that even creation was mortal, and only one in a cycle of creations that came and went from a timeless source, the fount of all becoming. The final step was to establish how, by observing closely the order of the universe, they might gain for themselves a return to this eternal source.

Thus was born what Eliade calls the "thirst to abolish dualism": "It existed at the most primitive stages, which indicates that man, from the time when he first realized his position in the universe, desired passionately, and tried to achieve concretely, a passing beyond his human status."[11]

---

*Michael Loewe, *Ways to Paradise* (London: George Allen & Unwin, 1979, 60): "Such a mirror presents the two discs [of the divination board] fixed rigidly and permanently in the most favourable position of the cosmos, in the hope that the deceased person with whom it was buried would be blissfully situated in that position for eternity."

Eliade suggests that this "thirst" is somehow fundamental to the human psyche. It brings us back to the question posed in the introduction to this book regarding the source of this thirst. Perhaps it is impossible to say what it is within the human mind that drives it to create order, to organize experience. Indeed, it seems to be the process that makes mind possible at all. Many might argue that it is the creation of organization that makes possible life itself. But I believe that we can now say how this "thirst to end dualism," the ultimate expression of the creation of order, first came to have the shape we know it in now. It derives from that long moment when we "first recognized" our "position in the universe," when the perfect world of the First Time came to an end and our ancestors set out on their journey through the machinery of time. From then on, we have carried the hope of finding our way back someday to that "deathless" world.

Interlude

SPIRITUAL
ARCHITECTURE

ECHOES OF ANCIENT MINDS

So profoundly in accord across all their geographical and chronological distribution do these stories seem to be that they can only have a common significance. As Santillana says, "[U]niversality is in itself a test when coupled with a firm design."

"When something found say, in China, turns up also in Babylonian astrological texts, then it must be assumed to be relevant, for it reveals a complex of uncommon images that nobody could claim had risen independently by spontaneous generation."[1] This argument gains even more weight when one adds such an unpredictable image as that of the fallen tree from areas as diverse as South America, Siberia, and Australia, or the world-embracing serpent from the Borneo rain forest, a Norse mountain, or an African village.

Uncommon these images are indeed, and their distribution can be far wider than Santillana envisaged. To my mind, no other interpretation of this universality comes anywhere close to the consistency of the "astronomical" one outlined here.

Such universality provokes a question that has not been addressed within the text: How did such universality come about? These are far from obvious ideas, as Santillana points out, unlikely to arise spontaneously in different cultures as part of the natural process of storytelling. That it has been possible to fit them all into a coherent and firm design makes such separate "spontaneous" origins even less likely. The complete design itself seems to be universal.

Only three answers are really possible. One is that these ideas were spread across the world by people who had the technical means to travel and the power to impress those they visited with their wisdom. Another is that they were first formulated in the very earliest times, before the dispersion of peoples around the globe. Neither of these suggestions conforms to our accepted notion of human history. Indeed, these accepted notions also rule out what is perhaps the most acceptable explanation, that these stories spread gradually across the world, as many of the most virile versions come from cultures considered both ancient and isolated, the Australian Aborigines being the most obvious.

In fact, we are unlikely to resolve this question within our present notions of prehistory. For now, it is one more item in the body of evidence that demands a serious reappraisal of the subject. Evidence continues to mount and continues to be discarded as "fanciful" by mainstream academic scholarship. Such questions will remain unanswerable until scholars of prehistory are prepared to reevaluate what material is available and to gather what is missing.

Equally challenging are the questions this interpretation presents about the structure that it discerns beneath the variety of mythical images, one that is in fundamental accord with the elements of spirituality as we understand them now and as they have persisted through the ages.

It seems to me that if we begin to grapple with the question of whether astronomy determined the shape of the spiritual or vice versa, we are actually missing the point. Instead of trying to establish "which came first," I would suggest that for the ancients (how easy it is to use that word, how much more difficult to actually give these people a real existence) there was no distinction. Astronomical observation and spiritual striving were the same process, and they are both manifestations of what Daniel Dennett describes as the "epistemological thirst" that he sees as one of the defining characteristics of all living organisms.

But humankind goes one step further; we want to belong, to find a place and a purpose. It is this kind of thirst that appears to have been born at that time when the ordered universe, in which humanity must have seemed to have a natural role to play, began to disintegrate.

"What is a solstice or an equinox?" asks Santillana. "It stands," he says, "for the capacity of coherence, deduction, imaginative intention and reconstruction with which we could hardly credit our forefathers. And yet there it was. Way back in time, before writing was even invented, it was measures and counting that provided the armature, the frame on which the rich texture of real myth was to grow."[2]

It was, Santillana suggests, "the effort of sorting out and identifying the only presences which totally eluded the action of our hands [that] led to those pure objects of contemplation, the stars in their courses. The effort at organising the cosmos took shape from the supernal presences, those alone which Thought might put in control of reality, those from which all arts took their meaning."[3]

And all this measuring and ordering, this construction of meaning, hangs on the notion of separation. Whether we read those stories that imply a perfectly created world before the parting of heaven and earth, before the "raising of the sky," or those that talk, as Plato does, of the crafting of time out of "brute matter," it is clear that identification of imperfection underpins the whole edifice.

Only by acknowledging the concept of a perfectly aligned harmonious creation can we appreciate the significance of the dilemma of the ancients. From their majestic solution to the problem of the decay of that perfection has arisen the philosophers' puzzle of the One and the Many, the spiritual seekers' thirst for reunification with the Godhead, and the physicists' search for the link between the Explicate and the Implicate Order.

It is a priceless insight that *Hamlet's Mill* offers, this vision of how a meaning might be discerned for the notions of heaven and earth and their separation, which enables us to see that all three are only aspects of the same search.

I do not want to suggest, however, that heaven and earth are nothing but the simple interpretations applied to them here. The process of mythmaking is itself one of creating a time-bound image of an eternal and all-embracing idea. To quote Santillana again:

> From whichever way one enters it one is caught in the same bewildering circular complexity, as in a labyrinth, for it has no deductive order in the abstract sense, but instead resembles an organism tightly closed in itself, or even better, a monumental "Art of Fugue."
>
> Stars, numbers, colours, plants, forms, verse, music, structures—a huge framework of connections is revealed at many levels [as if one's point of view were from within a kind of universal Mandelbrot set]. . . . One is inside an echoing manifold where everything responds and everything has a place and a time assigned to it. This is a true edifice, something like a mathematical matrix, a World-image that fits the many levels, and all of it kept in order by strict measure. It is measure that provides the countercheck. . . . When we speak of measures, it is always some form of Time that provides them.[4]

Cosmological time, that is, which is, after all, something "wholly other" than can be told by the hands on a clock. Again, Santillana makes the point:

> Cosmological time, the "dance of the stars" as Plato called it, was not a mere angular measure, an empty container, as it has now become, the container of so-called history; that is, of the frightful and meaningless surprises that people have resigned themselves to calling the *fait accompli*. It was felt to be potent enough to control events inflexibly, as it moulded them to its sequences in a cosmic manifold in which past and future called to each other, deep calling to deep. The awesome measure repeated and echoed the structure in many ways, gave Time the scansion, the inexorable decisions through which an instant "fell due." . . . These interlocking measures were endowed with such a transcendent dignity as to give a foundation to reality

that all of modern physics cannot achieve; for, unlike physics, they conveyed the first idea of "what it is to be." . . . Whatever idea man could form for himself, the consecrated event unfolding itself before him protected him from being the "dream of a shadow."[5]

From what we have discerned so far, we can go further. We can describe what this "foundation of reality" might be made up of, how the ancients described "what it is to be."

THE FOUNDATION OF REALITY

What do we mean by meaning? In the normal course of events, we read meaning into a message or experience by identifying what we might term *isomorphisms,* patterns that the message shares with all or some part of the rest of the world that we already know. That is to say, we regard it as containing elements that can be seen as encoding, in the form of metaphors, elements of some previously understood message. According to Julian Jaynes, "Understanding a thing is to arrive at a metaphor for that thing by substituting something more familiar to us. And the feeling of familiarity is the feeling of understanding."*

It has ever been the task of poets of one kind or another to point out or, indeed, to create these metaphors. (By poets, I mean all those whose lives involve exploring and elaborating the web of metaphor— the Greek word *poesis* means "creation.") By recognizing connections between things that had previously seemed unconnected, poets weave an ever more elaborate fabric of meaning in which each thread illuminates all the others. This recognition is based on the imposition of patterns taken from existing understanding onto new experiences. From such pattern making, organization emerges. Each new metaphor is thus an extension of the web that unites and orders experience.

---

*Julian Jaynes, *The Origin of Consciousness in the Breakdown of the Bicameral Mind* (Boston: Houghton Mifflin, 1990, 52). Much of the thinking in this section was inspired by Jaynes's idea of consciousness as "the work of lexical metaphor."

This fabric is a characteristic of mind—more or less elaborate according to the degree of superimposition of these metaphors—by which the world is understood. When we speak of the human spirit, I take it to refer to the arena in which this fabric is prepared. In this sense, spirit might be considered the "organ" of meaning, responding in degrees of "profundity" to the accord or resonance that can be perceived to exist between any new experience and the existing fabric of meaning.

Although it is infinitely enrichable, this fabric is woven on a primordial loom whose warp and weft were laid down at the very beginning of the human quest for meaning. Creation myth is a concise expression of the underlying pattern of this fabric. When written in astronomical rather than mythological language, the pattern is seen to be ultimately generated by that particular astronomical phenomenon known as the precession of the equinoxes.

Today the effects of this process go more or less unrecognized. There was a time, however, in the very distant past, when these effects formed the generative seed of meaning. The ancient stories are a description of the process. By exploring their contents and structures, it has been possible to uncover the original referents for all future "similarities," the "archetypal" signifiers, the first impressions made in the formless matter of understanding. They are, indeed, the architecture of the spirit.

LIVING IN THE MYTHICAL WORLD

Culturally separated histories have colored this "foundation of reality" in a multitude of ways across the world, emphasizing some aspects, diminishing others, so that today it is sometimes difficult to identify the commonalties and all too easy to exaggerate the differences.

In increasingly secularized societies, it is even difficult to acknowledge the existence of these underlying patterns, so remote have we become from their mythical embodiments. Instead, we choose to see our concepts and institutions as being the creations of our developing "civilization." We regard each separate field of endeavor as having its own determining principles. Once we begin to uncover the outlines of these archetypal patterns, however, it becomes possible to recognize, beneath the separate disciplines of philosophy and religion, of politics and science, of law and sociology, the original unifying principle of relationships.

For the mythmaking societies, it was the process of identifying or, rather, creating relationships that lay at the heart of their organization of the world around and within them. They appear to have gloried in the extension of simple relationships of all kinds, weaving ever more elaborate textures into the web of their understanding. All things and experiences that in some way manifested a particular characteristic were joined together in a relationship of shared imagery and potency.

Among the Dogon tribes of the Sudan, according to the anthropologist Griaule,

> everything can be regarded as a symbol. . . . And this arises out of the system of classification itself. Because an element in a "family" has a fixed place in a given series, its essential role can be summed up in the symbolic relationships it has with the other elements in the series, with the eponymous element, and via the latter with the rest of the family, which is itself hinged to the universal function. . . . Every mythical pattern is repeated in a series of instruments, things, institutions, gestures which diversify it. For instance, the ark . . . which came down from heaven to earth is reproduced in the Niger fisherman's canoe, the weaver's shuttle, the toll box, the anvil, the grain loft, the seats in the house, the drum and the harp-lute.[6]

To help clarify the processes whereby this kind of myth-guided world evolves, the anthropologist Lévi-Strauss coined the term "wild thought" (*la pensée sauvage*). It offers us a wonderful image of creative intelligence free to roam across domains untamed by the rigid concepts of our logic, which insist on the unassailable division between *either* and *or*. Wild thought embraces instead the notion of *both* and *and*. It then goes seeking out inspiration on this premise. From what it gleans, startlingly different models of the perceived universe can be constructed.

Lévi-Strauss explains how "mythical thought surpasses itself and contemplates, beyond images still clinging to concrete experience, a world of concepts defined no longer by reference to an external reality, but according to their own mutual affinities manifested in the architecture of the spirit."[7]

In remote areas of the world, societies organized around such apparently exotic worldviews still exist, or at least existed until quite recently. They are described in some detail in the reports of nineteenth- and twentieth-century anthropologists. These societies still tell mythical stories of their origins, stories that they place at the foundation of their lives and upon which they organize their whole social structure. In these societies, not only the mental systems—ethical, economic, theological, political, and so forth—but also social institutions, human relations, and even the built environment conform to the models laid down in the myths. Differentiation between the various activities, their disciplines of thought, and their intellectual constructs does not exist; they are all somehow encapsulated in the formulations of myth.

PRIME THOUGHT

In Africa it seems as if the whole of human life is contained within the mythical framework, as if the difference between the sacred and the profane no longer existed. Mythology provides man with models on which he must base his conduct, from the gesture of sowing seeds to the act of love, from house building to the touch of the fingers on the musical skin of the drum. . . . For the whole of Black Africa we can affirm the primordial importance of the myth, both as epistemology and as the basis of a theory of symbolic knowledge and as the basis of social, political, even economic structures, which are nothing more nor less than exemplifications of mythical patterns.[8]

This concept of myth as a vast and complex determinate structure for both social and spiritual life is one we may find difficult to comprehend. We are used to visiting the world of myth in the same way we visit a museum. Around us are excerpted exhibits, plundered from their own worlds and displayed for our amazement. Whether glorious or crude, they are presented to us as solemn reminders of the incomprehensible products of the human mind in a state of innocent ignorance. We may wonder at the extravagance and fertility of the invention, marvel at the impressive achievements, or smile at the uncouth nature of the more exotic tales, but the overriding impression we are led to receive is that here are the products of the childhood of understanding, things that we have put aside.

Even when we have been introduced to the works of the Jungians, with their attempts to describe myths as the "manifestations of the unconscious," or to those various theorists who have analyzed myths in terms of symbolism, of functionalism, of structuralism, or even of ecology, it comes as something of a shock to discover this world of mythical complexity and organization. Clearly this is a product of minds far removed from their infancy. Within this complexity, whole societies find their inspiration for intricate and sophisticated models of social structure, models that have proved sufficiently sustainable to survive apparently little changed across thousands of years.

Krupp describes, for instance, the world of the Desana Indians in equatorial Colombia: "The six corners of the tribal territory are marked by six waterfalls, each a place where the head of one of the six original giant anacondas meets another's tail. Each of these snakes stands for one of the six rivers that frame the traditional homelands."[9]

This hexagon of landmarks is the earthly equivalent of a "giant hexagon of stars, centred on the belt of Orion." The terrestrial hexagon is centered on the intersection between the Pira-Parana River and the earth's equator. Krupp explains: "Here where the sky is said to cohabit with the earth, is the place where Sun Father erected his shadowless staff and fertilized the earth."[10]

This spot is the whirlpool entrance to the womb of the earth, and from here the first people emerged at the beginning.

"Because of its importance as an organizing principle of thought, the hexagon metaphor reappears in one aspect of Desana tradition after another. All hexagonal shapes in nature have significance for them . . . even the shell of a particular land tortoise. Each cell in the shell's pattern of hexagons symbolizes a character in the creation myth or an organizational principle of society—the family, for example, or marriage into another family. Desana rules for marriage exchange are visualised in terms of a hexagon."[11]

Here are sky, earth, nature, culture, and myth united into one model of the universe, a model still actively inhabited by a tribe of equatorial Indians but built of images as ancient as any we know of. We would make a major step forward if we were to describe these earliest expressions of understanding as "prime" thought, rather than "primitive." We would then have in one word the notion of "first," together with that of "quality," but also with the sense of "an indivisible quantity." At least we should then be better prepared to face the full impact of myth, which, beneath the museum-exhibit surface of the populist imagery, is seething

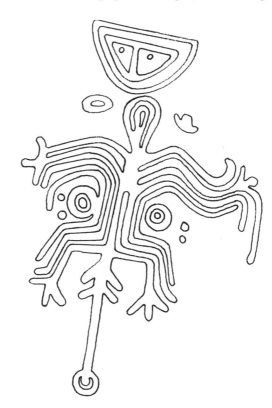

Fig. 26. Sun-Father Fertilizes the World. Detail from a petroglyph on the Rock of Ní, a sacred site of Colombia's Barasana Indians.

with an abundance of concepts so complex and obscure that it threatens to sweep us away on a tide of incomprehensibility.

Nor is this a characteristic only of those myths reported from remote and isolated contemporary societies. When we look more closely at the most ancient records, from Egypt, from Mesopotamia, from India and China, the myths of creation stand complete and almost incomprehensibly elaborate. All aspects of life seem to be embraced by their intricacy. Art, government, music, ritual, family relationships, architecture, even writing and the alphabet, appear as part of this completeness. Even a preliminary reconnaissance of the material reveals that, whatever else they are, myths are much more than the product of minds in a benumbed sense of uncomprehending fear before the forces of nature. They represent the products of highly developed intellects, revealing an immensely complex and profound awareness of that most fundamental of human endeavors, the art of organizing experience.

The ancient Hindu scriptures of the Rig Veda, considered among the earliest of sacred writings, reveal this complexity both in their language and in their structure. In her introduction to her translation, Wendy O'Flaherty discusses the formidable difficulties of making sense of such dense and paradoxical writings. One such difficulty, which she describes as a "form of deliberate confusion," is the use of

> mutually illuminating metaphors. Certain concerns recur throughout the Rig Veda . . . the themes of harnessing and unharnessing, which shift in their positive or negative value (sometimes good, sometimes bad); the closely related theme of finding open space and freedom . . . in contrast to being hemmed in or trapped. . . . These are linked to other constellations of images; conflict within the nuclear family . . . the preciousness of animals . . . the wish for knowledge and immortality. . . . The problem arises when one tries to determine which of these are in the foreground and which in the background of a particular hymn: are the cows symbolic of the sun or is the sun a metaphor for cows? The careless or greedy exegete finds himself in

danger of rampant Jungianism: everything is symbolic of everything else; each is a metaphor for all the others . . . when asked to pinpoint the central point of a verse, he will fall back upon the traditional catch-all of the short-answer questionnaire: "all of the above."[12]

In addition to its immense linguistic and semantic complexity, the Rig Veda reveals structural characteristics that are far from accidental. It is said to be composed of 10,800 verses, each of 40 syllables, making a total of 432,000 syllables in all. This is no arbitrary number. The fire altar, assiduously dismantled and rebuilt each year for the *agnicayana* ritual, contains 10,800 bricks, each one representing an individual part of the created universe. There are 108 classical Upanishads, and the same number, together with its factors, reappears throughout Indo-European myth and temple architecture.*

The Rig Veda is a concentrated expression of the immense web of relationships that the myth tellers created from their experience, from the messages they read in the world around them. It is this web of mutual affinities, this ever-increasing texture of organization, that generated the meaning upon which their society was constructed.

In weaving this web, the myth tellers were laying down the foundation for all the fundamental ideas about human spirituality and culture, of "ways of being in the world," from "fertility cults" and ancestor worship to Plato's doctrine of essences and the unity of all existence, from original sin and salvation to the concept of law and order, from domestic architecture and family relationships to immortality and the eternal godhead.

Each of these ideas is linked to the others, not in any nebulous way but according to a fundamental set of principles, expressed in the myths of creation and organization. Behind it all lies the original link made between the pattern of life on earth and the patterns displayed in the

---

*The significance of the numbers 54, 108, and 432 (in various multiples of 10) has often been noted but never fully explained. They are all, however, factors of the complete precessional number 25,920.

night sky. It was an ancient convention, but it survived more or less intact in the work of artists and natural philosophers up to (and including) Sir Isaac Newton, to presume that the whole of "nature" conformed with the mode of existence of the stars. Indeed, as one might expect, in the light of what has been said so far, Mother Nature herself had decidedly celestial origins.

The Myth of Reunion

THE MOTHER OF ALL

EVERLASTING PLACE

So far, we have allowed only heaven and earth as elements into our cosmos, discerning in their separation and its consequences the origins of the philosophers' notions of Being and Becoming. In discussing the manner in which the world of becoming arises out of the timelessness of being, however, Plato makes several obscure (or obscured, if Plato is deliberately avoiding being specific) references to a third "substance," to which he gives a number of different titles; the first of these is Place: "Being, Place and Becoming," he writes, "were existing, three distinct things, even before the Heavens came into existence."[1]

He describes this "Place" in slightly more detail as "accommodation" for the becoming: "One kind is self-identical form, ungenerated and indestructible, being the object which it is the province of Reason to contemplate; and a second kind is that which is named after the former, apprehensible by Opinion; and a third kind is everlasting place, which admits not of destruction, and provides room for all things that have birth."[2]

To describe this third kind merely as a "womb," as some have done, would be misleading, however, for Plato also calls it "Nurse and Recipient." It is clear that he is unable or does not want to be more specific. "Place," he says, "is apprehensible by a kind of bastard reasoning, by the act of non-sensation, barely an object of belief."

This intriguing "substance," which is both "baffling and obscure" and "the nurse of all becoming," seems to introduce a division of the idea of the eternal being into two parts, imagined as male and female. It is this duality of the perfect that Plato must have found baffling, although he was aware of the concept; he tells us, "The recipient is fittingly likened to a Mother, the model of becoming to a Father and the nature that arises between them to offspring."[3]

This third substance that is causing Plato his bafflement is nothing less than the "Great Goddess," the Mother of All. Valentinus was less confused and was content to describe the divine as a dyad consisting, in one part, of the ineffable, the Depth, the primal Father, and in the other of Grace, Silence, the Mother of All, seeing these two as "parents" of the "emanations." According to his notion, Silence receives the seed of the ineffable Source; from this she brings forth all the emanations of divine being, ranged in harmonious pairs of masculine and feminine energies.

This procreational image of creation is subtly reversed in other descriptions, however. Instead of the Silence receiving the seed, it is the Maker who receives his instructions; in an Orphic fragment we learn, "The Maker of All is said to have gone into the oracle of Night before the whole creation and from there to have been filled with the divine plan and to have learned the principles of creation."[4]

From this point of view, Silence is the plan according to which creation is to proceed. In fact, Valentinus considered this Silence to have been in existence before the dyad: "From the power of Silence appeared a great power, the mind of the universe, which manages all things and is male, and a great intelligence, a female which produces all things."[5]

This notion of the plan for creation appears later in medieval natural philosophy as *sapienta Dei,* the wisdom of God, or the *archetypus mundus,* the "exemplar" of the universe. According to Marie-Louise von Franz, "It denoted the timeless, pre-existent cosmic plan or antecedent world-model, potential in God's mind; according to which he realized actual creation."[6]

The source of this image is the "Wisdom" of the Proverbs of the Hebrew scriptures, whose nature is revealed by her various proclamations:

I was set up from everlasting from the beginning or ever the earth was. When he prepared the heavens I was there (Proverbs 8:23).

The lord by wisdom hath founded the earth; by understanding hath he established the heavens (Proverbs 3:19).

I came forth from the mouth of the Most High and as a mist I covered the earth / In the high places did I fix my abode and my throne was in the pillar of cloud / Alone I compassed the circuit of heaven and in the roots of Tehom [the Depths] I walked / Over the waves of the sea and over all the earth / And over every people and nation I held sway. (Book of Sirach 24)[7]

The Greeks knew of a similar tradition referring to Wisdom as "Intelligence," to whom Orpheus gave the name Moira ("degree"): "Even before Zeus was mentioned Moira, the intelligence of the god, existed always and everywhere."[8]

Since Plato himself confessed to finding this matter "baffling and obscure," we are hardly likely to be able to do better. It is possible, however, to propose two ways of employing the One and Many model of the creations to interpret this material.

First, where we read about the plan in God's mind, it seems most likely that we are talking about that "one God" who is the first of many, what the Hindu texts call an *avatar,* rather than the One God who is the eternal completeness. In this sense, "everlasting place" represents the starry sky already divided into the creations that will come. The act of the creator in working with this plan is to put into operation the appropriate manifestation of the plan.

The second way of understanding these concepts is to see Place as the recipient, as Plato has it—the starry sphere in its entirety, awaiting the imprint of the current archetype, selected by the creator from the ineffable depth, the entire collection of potentialities. It is as if in this sense, the depth represents the Frame of Time independent of any particular fastening to the stars, like a three-dimensional block of typeface (the archetype) waiting to be turned to the correct position by the creator before being impressed onto the spherical "page" of the recipient.

This second interpretation seems to be what Valentinus was describing and the one that Plato described most fully, although he does seem to be struggling here to combine the two: "If we describe her as a kind invisible and unshaped, all receptive and in some most perplexing and most baffling way partaking of the intelligible we shall describe her truly."[9]

THE QUEEN OF HEAVEN

Little remains in our common understanding of this all-encompassing Wisdom Goddess who participated in the creation. We have come to accept the maternal, "recipient" nature of the female deity; we have made her Mother Earth and Mother Nature. We have, it is true, kept some remnant of her disposing power in the blind image of impartial Justice.* But we have lost sight of the real significance of the "well-known fact" that Eliade points to in discussing the multitude of examples of the origin of social order: "The fact that human justice which is founded upon the idea of 'law' has a celestial or transcendent model in the cosmic norms is too well known for us to insist upon it."[10]

It is to these "cosmic norms" that we must look to understand Plato's notion of everlasting place. She is, in fact, Queen of Heaven, as Merlin Stone pointed out: "Descriptions of the female deity as creator of the universe, inventor or provider of culture were often given only a line or two, if mentioned at all; scholars quickly disposed of these aspects of the female deity as hardly worth discussing. And despite the fact that the title of the goddess in most historical documents of the Near East

---

*Stephen Langdon, speaking of the goddess Ininni (or, more commonly, Inanna) of Sumeria, calls her "The Divine Mother Who Reveals the Laws" (Merlin Stone, *Paradise Papers*, 216).

was the Queen of Heaven, some writers were willing to know Her only as the eternal 'Earth Mother.'"[11]

Stone provides numerous examples of other names for this goddess, including "Lady of the High Place," "Celestial Ruler," and "Lady of the Universe." She is frequently referred to in connection with oracles, and Stone comments: "It was not only the belief that the priestesses could see into the future that made oracular divination so popular, but the idea that these women were understood to be in direct communication with the deity who possessed the wisdom of the universe."[12]

This is confirmed by the Book of Wisdom: "[Wisdom] has fore-knowledge of signs and wonders, of the unfolding of the Ages and of the times."[13] The wisdom of the universe is the knowledge of how it will proceed. Homer tells us that Tethys (whose name is derived from *tithenai,* "to dispose"), together with the serpentlike Oceanus, who girdles the world, gave birth to all the gods. The Greeks knew several deities with this kind of attribute: Tethys, Gaia, Themis, Metis. All represented the power to generate and order the universe, as Detienne and Vernant explain in their *Cunning Intelligence in Greek Culture and Society:* "[Metis's] gift of metamorphosis fitted her to represent the complete unfolding of a cycle of forms which is, in a sense, already contained in the original form."[14]

But perhaps the fullest descriptions of this goddess come from Egypt and India. We have already introduced the concept of "cosmic law" (see "Out of the Garden"). In Egypt, this concept was embodied in Ma'at, whom Wallis Budge describes as "the goddess of absolute regularity and order, of moral rectitude and of right and truth. She assisted [Thoth] at the work of creation," he adds.[15]

As goddess of truth and moral order, Ma'at appears at the weighing in the balance of the heart of the dead, and in this role she has become regarded as a figuration of Justice, but, as Stechinni explains, she is more properly understood as "the proper cosmic order at the time of its establishment."[16]

In Hindu tradition this "transcendent model" is Rita (from the verb *rta*—"to fit"), which, according to the *Dictionary of Hinduism,* "in ontology represents the immanent dynamic order or inner balance of the cosmic manifestations."[17]

To clarify this description, it is worth recalling the words of Eliade regarding the role of Rita, who, he says, "designates the order of the world, an order that is at once cosmic, liturgical and moral. The creation is proclaimed to have been effected in conformity with Rita; it is repeatedly said that the gods act according to Rita, that Rita rules both the cosmic rhythms and the moral code."[18]

THE HEARTH OF THE UNIVERSE

Merlin Stone emphasizes the extent to which this "Queen of Heaven" is referred to as a serpent, whether in her benign form as oracle or her monstrous form as destroyer. This serpent, which sometimes is the goddess, sometimes her consort or her son, appears with her at the creation in many stories; the Egyptian goddess Hathor in one text is described as the Great Serpent who created the earth and who, in her anger, decides to destroy it.

"In several Sumerian tablets," writes Stone, " the goddess was simply called Great Mother Serpent of Heaven."[19] In Egypt she was Uatchit, "the Lady of Flames . . . the eye of Ra,"[20] who, in the Egyptian Book of the Dead, is said to "make" or "work" right and truth: "right and truth which the goddess Uatchit worketh."[21]

This goddess is clearly something more concise than an image of the starry sky. Her knowledge of the future of the universe, of the orders of creation, suggests that she should be seen as the complete cycle of the ecliptic stars. She has a further aspect, however, which, while it may at first seem incongruous, will lead us to a more profound understanding.

The Hindu goddess Svasti is regarded as the "goddess of the home and prosperity." That her original domain is more profound can be deduced from her ancient symbol, the swastika, said to represent "cosmic procession and evolution around a fixed centre."[22]

The prosperity that she guards is that of the entire cosmos. Her home is the home of the gods, as Plato explains in his description of her Greek equivalent, Hestia: "Zeus, the mighty lord, holding the reins of a winged chariot, leads the way in heaven, ordering all and taking care of all; and there follows him the array of gods and demigods, marshalled in eleven bands; Hestia alone abides at home in the house of heaven; of the rest, they who are reckoned among the princely twelve march in their appointed order."[23] That is to say, the twelve gods of the host of heaven, the zodiacal constellations, circle around "the house of the gods," but the goddess remains at their center, tending the "hearth of heaven."

This hearth, the "fixed center" around which the cosmic procession rotates, can only be the ecliptic pole, the only element in the cosmos deserving of Plato's title "everlasting place." Here the goddess resides and presides over the course of the creations. The Pythagorean tradition retained the full import of the relationship between the goddess and the central hearth. According to the Greek philosopher Aetus, "Philolaus the Pythagorean says that the fire is at the centre calling it the hearth of the universe; he calls it the hearth of the whole, the mother of the gods, the altar and substance and measure of nature."[24]

If this hearth of the universe is the ecliptic pole, then the "Great Mother Serpent of Heaven" must be the stars of the constellation Draco, the dragon that lies curled around the pole; it also explains her title "Lady of the Flames."

Crucial to understanding the manner in which this central "fire" is able to generate "the substance and measure of nature" is the symbol of the swastika, for it represents the "fire-drill," the ancient fire-creating device consisting of two sticks, one that rotated vertically in a small hole in the other. Either by rubbing it between the hands or by the use of a bow whose string was twisted around it, the vertical stick was twirled vigorously until sufficient heat was generated to ignite the kindling placed in and around the hole in which it turned.

Igniting fire by this means was analogous to the act of creation, for it was just such a twirling motion that once generated an entire universe, according to the description in the Mahabharata and as depicted repeatedly in Hindu monuments and paintings. It was during a battle with the demons that the "celestial treasures" fell into the Milky Ocean. Here is the description in the Mahabharata, where it is referred to as the *Amrita manthana,* "the Churning of Immortality": "Let the gods and demons churn the ocean which is like a churning pot. . . . Churn the ocean, O gods, and you will find ambrosia [*amrita,* the drink of immortality]."[25]

The great serpent Ananta ("Endless") then uproots the mountain Mandara, and the gods take it to the ocean. "We will churn your waters," they say. For a share in the prize, the Lord of Waters agrees to bear the intense agitation from the whirling of Mandara. Vishnu's avatar Kurma the tortoise then agrees to carry the tip of Mandara on his back, and, using the serpent Vasuki as "the cord," they begin to churn:

Fig. 27. The Churning of the Milky Ocean. Even in such a simplified image of this momentous event, the "demon" (at right) is distinguished by its hair and horns (see "The Wild and the Tame," page 153).

"The gods acted together with the demons, for they all wished for the ambrosia."[26]

First to emerge are what appear to be the constellations, one by one; at last the physician of the gods emerges carrying the bowl containing amrita. When the demons seize the bowl, they are deceived by the great god Narayana into giving it up; Narayana then gives it to the gods, who drink enthusiastically before yet another vast battle begins between them and the infuriated demons. Santillana tells of a traditional version from the north in which the alternating motion of the fire-drill is replaced with that of the butter churn: "In the myths of the Kalmucks . . . the world came into being when four powerful gods got hold of Mount Sumeru and whirled it around in the primordial sea, just as a Kalmuk woman turns the churning stick when preparing butter."[27]

It was *Hamlet's Mill* that first suggested that this story is a description of the precessional process. The authors point to similar images of cooperation between the rivals Horus and Seth in Egypt, who are shown holding opposite ends of a rope with which they are rotating some kind of churn. We might also recall the Japanese story of the emergence of the "celestial pillar" as a result of the stirring of the endless sea (see again "The First Earth," page 36).

Two aspects of the story are important if we are to appreciate the justification for this suggestion. First, it requires the cooperation of two opposing forces, and second, like both the fire-drill and the churn from which its generative imagery is variously derived, it involves an alternating motion. Together these two elements seem able to maintain the whole process of creation and destruction. The fundamental cultural act of fire-lighting represents—that is to say, re-presents, or re-enacts—the process of cosmic creation.

If the Mahabharata's image of the Milky Ocean churning seems too fanciful to carry such weight, we can turn our attention to what Plato has to say about alternate motion in the universe: "There is a time when God himself guides and helps to roll the world in its course; and there is a time, on the completion of a certain cycle, when he lets go, and the world, being a living creature, and having originally received intelligence from its author and creator, turns about and by an inherent necessity revolves in the opposite direction."[28]

Plato is retelling the ancient tale of how, at the end of the age of Kronos, God let go of the world and it turned back on itself. At the end of this age, God will take hold of it again:

> When the world was let go, at first all proceeded well enough; but, as time went there was more and more forgetting, and the old discord again held sway and burst forth in full glory; ... and there was a danger of universal ruin to the world. Wherefore God, the orderer of all, in his tender care, seeing that the world was in great straits, and fearing that all might be dissolved in the storm and disappear in infinite chaos, again seated himself at the helm; and bringing back the elements which had fallen into dissolution and disorder to the motion which had prevailed under his dispensation, he set them in order and restored them, and made the world imperishable and immortal.[29]

This "reversal" sometimes comes disguised as "darkness." Thus, at the crucifixion of Christ "there was darkness over the whole land."* This "darkness" appears to be related to the notion that the end of a heavenly cycle would be marked by an eclipse. Hence Cicero, in his *Dream of Scipio,* says: "For as the sun in old time was eclipsed, and seemed to be extinguished, at the time when the soul of Romulus penetrated into these eternal mansions, so, when all the constellations and stars shall revert to their primary position, and the sun shall at the same point and time be again eclipsed, then you may consider that the grand year is completed."†

---

*Gospel of Mark 15:33. David Ulansey says of the "veil of the temple," which, according to Mark (15:38), was torn in two from top to bottom at the crucifixion, "It typified the universe. ... Portrayed on this tapestry was a panorama of the entire heavens" (*The Heavenly Veil Torn,* at www.well.com/user/davidu/veil.html). The image of the sun "standing still" or even "going backward" appears several times in the Hebrew scriptures.

†Tullius Cicero, *On the Republic,* Book VI, 22 (www.fordham.edu/halsall/ancient/cicero-republic6.html). "Be assured, however," Scipio is told, "that the twentieth part of it is not yet elapsed."

Here, perhaps, we can begin to see how this alternating motion is related to the notion of the precession. First there is the motion toward disorder, the gradual movement of the precession pushing the Frame of Time away from its alignment with at first the Milky Way and later any one of the colures that mark the beginning of a zodiac sign. This motion is said to be generated by "the demons." Second is the motion generated by the gods, which pulls the Frame of Time into a new alignment with a new zodiacal colure, and thus marks the onset of a new period or era. It is at such a "moment" that the "sky" may be said to stand still, or even to travel backward, as the Frame of Time is "pulled" in the opposite direction to establish a new era of measurement.

As far as the Frame of Time constructed for the Age of Aries was concerned, the vernal equinox occurred as the sun arrived at the first point of Aries. Astrologers today still work with such a frame. By the time of the crucifixion, the actual position of the equinox coincided with the arrival of the sun at the first point of Pisces. The calendar of the Age of Aries was hopelessly awry. To put matters right, the "heavenly bodies" (the planets, including the sun) had to be held in position while the stars of the zodiac continued on their way. This is a metaphorical way of saying that the whole Frame of Time had to be dragged "backward" through the stars, against the usual motion of the planets (from west to east during the year) to arrive at a new "fixed point" in alignment with the new equinoctial colure. The gods would then let go of the universe, and, true to form, and as a result of the slow circling of the polar axis, it would "wind down," the forces of chaos would intrude, and disorder would return the world "into the infinite sea of dissimilarity."*

This is the stage we have now reached as the Age of Pisces draws to a close and the gods prepare to "wind up" creation once more, to align it with the equinoctial colure in Aquarius.

---

*An illuminating example of this tradition is included in Paul Isaac Hershon's *The Pentateuch According to the Talmud* (London: Samuel Bagster, 1883): "[T]he wise men of Israel maintain that the celestial wheel is fixed and that the signs of the zodiac attend the motion of the sun (i.e., each moves within the wheel part of the sun's course till relieved by the next sign, when it returns to its former position)."

NATURE AND CULTURE

THE WILD AND THE TAME

Whatever their origins (and we know that the Hindu demons, the Asuras, were originally devout and knew the Vedas, before they grew proud and overbearing), the two opposing elements, by whose temporary cooperation the universal fire was first drilled, can be regarded as representing the forces of order, on the one hand, and chaos—that is, order gone astray—on the other. The anthropologist Claude Lévi-Strauss described their interaction as "an opposition which is deemed essential in mythic thought and is perhaps the root-source of other religious images, including those of our own culture."[1]

For Lévi-Strauss, these forces represent the opposing concepts of nature and culture. The bringing into play of their opposition "either through the destruction of an original harmony or through the introduction of differential gaps which impair that harmony" results in humankind's access to culture being accompanied "on the level of nature by a form of deterioration entailing a transition from the continuous to the discrete."[2]

The confusion of the idea of nature as a unity (as opposed to the dividedness of culture) with that of nature as chaos (as opposed to culture as order) is as old as any. It can be equated with the confusion between the paradise of the First Creation (as opposed to the later, less perfect creations) and that of the eternal heaven (as opposed to the time-bound worlds). It seems an inevitable consequence of the process of separating out unity, and thereby generating dividedness, that the separation becomes one between the organized and the "unorganized." This "unorganized" is then regarded as a threat, and the image of "impending chaos" is born.

The conflict between chaos and order has been perhaps the most influential of all concepts in determining models for human thought, but it is not as inevitable as might at first appear. To be sure, the generation of order is a fundamental characteristic of consciousness itself; it seems that part of the way the human senses function is by creating divisions in "reality." We might regard chaos, then, as a residual awareness of the "dividing" that perception entails, in which case, it is chaos of the primeval sort. Order generated in this consequential way ought to be inviolate, a statement of permanent as-is-ness, separated from chaos in the same way we are separated from eternity. The idea that order is vulnerable—that is, threatened by chaos and prone to decay—is a consequence of all the magnificent collating of experience into models of interrelatedness that we have been exploring.

From this point of view, order characterizes the world inhabited and organized by humanity; chaos represents the powers that gather on the borderlands of human culture and continually threaten to disrupt it. In some African traditions, for instance, the village is regarded as an enclave of order, in contrast to the "bush," which is the harborer of all those forces, forever living outside the boundaries of "civilized" life, that threaten the security of that life. They embody the threat of the "wild" as opposed to the cultivated organization of the "tame." (One should not take this kind of tradition as necessarily representing a fixed attitude. In fact, in the same traditions, all the occupants of the "bush" can be seen to have their allotted places in the organized structures. These systems of classification are metaphors; they can run concurrently, contradictory but not mutually exclusive.)

It seems likely that from the earliest times, the forces of order and of chaos, the gods and the demons, were identified with notions of the tame and the wild. The medieval image of the Wild Man is derived from this concept; the wild forces are repeatedly described as having two distinguishing characteristics: they are hairy and they have horns (see fig. 27, page 149).

Thus, in the Rig Veda, Indra is said to be "king of the horned and the tame"; a medieval alchemical text calls upon us to put aside the "horns of pride" and the "hair of worldly superfluity."[3] These horns go at least as far back as the earliest Egyptian depictions of the deity Seth, the rival twin of Horus, shown as an unidentified beast, but distinguished by his square-ended, upright "ears";* later Hindu representations of the Asuras are similarly endowed with small but persistent horns. As for the hair, it is at least as old as the story of Gilgamesh, whose "twin brother," Enkidu, has grown up, wild and hairy, out in the badlands. Anyone familiar with the Hebrew scriptures will also call to mind the story of Jacob and his older, hairy twin Esau. The details of their birth, however, may have been forgotten: "There were twins in her womb. And the first came out red, all over like an hairy garment. . . . And after that came his brother out, and his hand took hold on Esau's heel."[4]

"That," says the commentary, "marks the end of one era and the beginning of the next."

THE ORIGIN OF FIRE

Lévi-Strauss unhesitatingly connected the transition from the state of nature to the constitution of society with one critical event: "This event, crucial for the life of mankind, is the theft of fire from the sky by a terrestrial hero."[5] He goes on to recount how this event, which made possible the organizing of the natural world and the erection of definitions and structure, was accompanied by the arrival of mortality:

---

*These "ears" reappear throughout myth and fairy tale, often as "asses' ears": the Persian deity Zurvan was sometimes said to have had them; King Mark, in the British tale of Tristan and Isolde, has them; Bottom in *A Midsummer Night's Dream* has them.

"In the temporal dimension, [the transition from the continuous to the discrete] interrupts the demographic flow and, by dividing it into generation levels, introduces differential gaps comparable to the differences between the animal species in the sense that what might have remained an undifferentiated fabric is fragmented into separate entities."[6]

The implication here is that it is the arrival of fire that marks what Christians might call the "fall from Grace." Although Lévi-Strauss never explicitly makes the connection, it is clear that this fragmentation of the undifferentiated can be equated with the separation of heaven and earth. Santillana and von Dechend confirm this equation by quoting the Mongolian nuptial prayer that says, "Fire was born when Heaven and Earth separated."[7]*

In fact, we can link this story with that other decisive event, the felling of the sky-tree. In his work *Myths of the Origin of Fire,* Frazer recounts the story that tells how "men and women used to climb up the great pine tree that reaches the sky; a hawk discovered the secret of fire sticks but in an argument burnt the tree. The people left in the sky became stars."[8]

Santillana and von Dechend confidently assign the notion of "fire" to the role of the equinoctial colure, quoting, among many other sources, the Rig Veda hymn to Agni, the sacrificial fire, which says, "Like the felly the spokes, so you surround the earth."[9]

As if to confirm the relationship between the original fire and the Milky Way, Frazer records the relationship between the arrival of fire and the two bright stars of the constellation Gemini, which stand on the east bank of the heavenly river. Castor and Pollux, he says, are "the two black fellows who brought fire to Tasmania. . . . They threw fire like a star from where they stood on the summit of a hill."[10]

---

*According to E. Lot-Falck ("Siberia: The Three Worlds," in *Larousse World Mythology*), this Fire is "Mother Ot" the Fire Queen whose warmth "produces and maintains existence." She also appears as goddess of the domestic hearth. Lévi-Strauss tells a similar story from New Guinea, in which cooking fire is received in exchange for the communication that had existed between the earth and the sky (*An Introduction to a Science of Mythology* 4, London: Cape, 1978, 622).

Of all those figures responsible for bringing fire to earth, the most renowned is surely Prometheus. As told by Hesiod, the story describes how Zeus, dismayed by humankind's increasing powers, had wanted to destroy it. Prometheus (who some said had created mankind in the first place) had dissuaded him, but Zeus, angered by a trick Prometheus had played on him, decided to withhold fire from humankind. Prometheus then stole the "far-seen gleam of unwearying fire" in a hollow fennel stem and delivered it to humankind.[11]

There seems little in this story to relate it to Lévi-Strauss's formative and divisive fire—not, that is, till we read Santillana's description of the dispute that rages over the very name Prometheus. He leaves little room for doubt that it was derived from the same root as that of both Mount Mandara and the Amrita Manthara (the Churning of the Milky Ocean): the Indo-European root *manth/math,* which implies "whirling or rubbing, as in a churn or a fire-drill"; Prometheus *is* the fire-drill.*

It takes a good deal of getting used to, this idea that such familiar elements of myth can suddenly take on such profound and precise roles. We are used to consigning "fire-worshippers" and "sacrificial fires" to the category of "primitive" religions. To be suddenly faced with the serious proposition that these images be regarded as representing concepts as sophisticated and precisely observed as these is a major challenge to our understanding. Nevertheless, the evidence mounts up relentlessly, once we know where to look.

In their book *Celtic Heritage,* Alwyn Rees and Brinley Rees describe the arrival of fire in Ireland: "It was at Uisnech that Mide, chief druid of the people of Nemed, lit the first fire. . . . From that fire were kindled every chief fire of Ireland,"[12] and they go on to relate this to the arrival of fire described in the Rig Veda, where it marks the beginning of time:

"In the Rig Veda the arrival of the Five Kindreds and the Eight Adityas marks the birthday of the Universe, the origin of months and days . . . 'the Spring-tide of the Cosmic Year, when dawn first shone

---

*This etymology is quoted by Santillana and von Dechend but is disputed by others.

for Man.' On landing, their first concern is to establish on earth a ritual in 'imitation of the First Sacrifice.' They erect a fire altar, and by this sacrificial act they gain possession of the land and secure their legal title to it."[13]

THE FIERY WELL

The immense Hindu "fire sacrifice" known as the Agnicayana involves the ritual destruction and re-creation of a complex altar (built of 10,800 bricks, a number whose significance has often been noted), which represents the organized cosmos itself. Its dismantling and reassembly mark the ending of one age and the beginning of the next, re-presented as an annual event. Although the structure and the content of this ritual are intricately and precisely defined in the scriptures, echoes of similar ideas can be heard in the many fire rituals connected with New Year celebrations. These frequently involve extinguishing all fires and relighting them from a central fire either kept permanently alight or reignited ritually (often using the fire-drill, the technological equivalent of the cosmic mill). The Roman Vestal fire, dedicated to Vesta (the Roman equivalent of Swasti), from which domestic fires were reignited after periods of mourning, is one such, but there are many other examples of the same notion.

Even today, we retain the concept of "housewarming" to establish occupancy of a new home. Ideally, this event should include lighting a new fire in the hearth; it represents in miniature the beginning of a new era.

In discussing the significance of the Hindu "fire-god" Agni, von Dechend and Santillana stress the point that there have been many Agnis known to Hindu myth, each one representing a new fire generated to mark the delineation of a new age—that is to say, in their opinion, a new location for the equinoctial colure. The first fire, however decisive its arrival for the future of creation, was destined to die with the age it created.

It is in this context that we must view the stories that tell of the theft of fire from heaven, stories told over and over around the world. Sometimes, as Santillana reports, it is a stag that carries out the theft, sometimes it is a bird; and in these cases, the bird is burned in the process and carries the mark on its feathers thereafter. Frazer describes these birds as wrens, although the species varies from country to country. In northern Europe, especially in the British Isles and Brittany, it is the goldcrest wren (not, as is so often repeated, *Troglodytus troglodytus* but rather *Regulus regulus,* the "King of the Birds").

In other stories, retrieving the new fire is accomplished by a "culture hero"; in fact, it is the possession of the creative fire that enables the hero to bring "culture" to his people. Nor is this "civilizing" process a once-and-for-all event; it must be repeated for each age with a new hero and a new journey to the underworld. In his book *The Hero with a Thousand Faces,* Joseph Campbell retells a story from the west of Ireland that describes the visit of the Prince of the Lonesome Isles to the Lady of Tubber Tintye ("The Well of Fire").[14] After passing all sorts of monsters, he reaches her palace, where he finds her sleeping on a revolving couch, the fiery well with its golden cover at her feet. He sleeps with her for six nights and eats from her inexhaustible table. When he leaves, he takes away with him three bottles of the water from the fiery well.

Behind this story lies a whole body of material concerning the visits of various heroes to the underworld to bring back a boon of some kind, but the Prince of the Lonesome Isle will have to stand for them all. A well of fiery water is both a potent image in itself and a fine example of the way in which complex mythical imagery is distilled into the matter of fairy tale. Campbell says of this story, "The well is the world navel . . . the bed going round and round being the World axis."[15] Of the fiery water, he says only that it represents "the indestructible essence of existence."

THE SAVIOR
OF THE WORLD

THE PRINCE OF GLORY

In traveling to the palace of the Queen of Tubber Tyntye, our hero has indeed reached the "navel of the world," for it is none other than the Queen of the Seas he is visiting. The lady has her throne at no other place than "the Confluence of the Rivers," the celestial south, where the two "branches" of the Milky Way are said to meet. Her fiery well is known in Hindu tradition as *svarnara*, the "celestial spring," the same svarnara to which Varuna fastened the sky and which is referred to as "the seat of Rita," the source of the laws of creation.

Santillana and von Dechend make it clear that this svarnara is to be identified as the source of the mysterious quality that Iranian tradition refers to as Hvarna, or *farr*.[1] According to the Persian epic *Shahnameh*, it was possession of this "royal *far*" that enabled the hero Hushang to bring prosperity into the world, including the craft of the smith and irrigation of the land. Similarly his successor, the emperor Jamshid, declares: "I am endowed with the divine Farr and I shall guide the souls towards light";[2] and it was by virtue of this farr that he brought all the arts to his people.

This mysterious quality has been translated as "glory," and as the "royal mandate," and is frequently pictured as a black bird.* In Babylon it is *melammu,* the blazing brilliance with which Ea endowed his offspring Marduk before sending him to confront the dragon Tiamat. In India it was regarded as being in the gift of Varuni, "the Waters," sometimes described as the "wives of Varuna" and the "goddess of intoxication." Eventually the concept became personified as Sovereignty or Fortune, and by Roman times it had become a goddess ("Fortuna") in its own right. Today it survives in the form of "Lady Luck."

SOVEREIGNTY

That the "mandate to rule" was conferred along with a draft from the celestial fiery well is confirmed repeatedly in the annals of Ireland. Queen Medb was the wife of nine kings of Ireland in succession, and, we are told, "[s]he would not allow a king in Tara without his having her for wife." It seems that she had this authority as a result of her access to the "Ale of Cuala."†

It must have been Queen Medb in one form or another who appeared to Niall as a black hag when he and his brothers were in search of water. She will give them water from the well she guards only in exchange for a kiss, which all except Niall refuse. In response to his embrace, however,

---

*It is difficult not to be reminded of the royal ravens that guard the Tower of London. The legend holds that the royal household is secure only as long as they remain there, and a small regiment of soldiers is maintained to serve them.

All this lends a very different aspect to Campbell's tale of the Prince of the Lonesome Isle. According to the Vishnu Purana, the "Confluence of the Rivers," where the Queen of the Seas has her palace, is the source of all the measures that determine the creation, for here "dwell the Great Rishis, after whose injunctions creation commenced; for as the worlds are destroyed and renewed, they institute new rules of conduct, and re-establish the interrupted ritual of the Vedas."[3] When our hero returns home with the flaming water from this source, he comes empowered to claim his kingdom and reestablish it according to these new rules and rituals; he might indeed be referred to as the Prince of Glory.

†"No-one will be king of Ireland unless the Ale of Cuala comes to him" (Alwyn Rees and Brinley Rees, *Celtic Heritage,* 75). Medb's name is related to the English "mead," Hindu *medhu*—"honey." Both these names have their root in the word that recurs so often in this study, *me,* meaning "measure."

she becomes "the most beautiful woman in the world." When Niall asks "What art thou?" she replies, "King of Tara, I am Sovereignty." Referring to the water she has offered him, she says, "Smooth will be thy draught from the royal horn, 'twill be mead, 'twill be honey, 'twill be strong ale."[4]

Similarly, when Conn has his vision of Sovereignty, she is serving the future kings of Ireland from a cup that she also presents to Conn himself; its contents are called *derg fluath,* which is both "red ale" and "red sovereignty."

This Irish "Sovereignty" has been identified with the Indian goddess Laksmi, who is the "personification of the right to rule." She is the consort of Indra, having wooed him with the drink of immortal power, Soma, the drink of which "none tastes who dwells on earth." Its power lies in the fact that the one who "knows its secret" has the ability to destroy demons and thus to win the rulership of the universe. It is the knowledge of the secret of Soma and the mandate of Agni, the ritual fire, that enables Indra to do exactly that.

THE WEAPONS OF GOD

To be sure of success, the hero who intends to redeem the creation from the clutches of the monster of chaos must come to the battle properly equipped. Like Indra, he requires the intoxicating draft from the Queen of the Waters. In addition, he requires one or more weapons, many of which take the most unlikely forms. To help him defeat the tyrant Zahak, according to the *Shahnameh,* the hero Farudin called for "cunning smiths to fashion me a heavy mace" and "[he] outlined in the dust a figure portraying the likeness of a buffalo's head."[5]

As Santillana points out, if this reminds us of Samson's "jawbone of an ass," it will not be in poor company.

To keep us sighted firmly on the cosmic path, however, we should add to our list the "mace" known as "Gadagada," which Vishnu made from the bones of the monster Gada, which he had slain. This mace "symbolizes invincible power or knowledge" and is the "power that ensures conformity with universal law . . . 'time the destroyer.'"[6]

Perhaps the most familiar of these demon-slaying weapons is the thunderbolt, a symbol of royal authority in Assyria and India but best known as the weapon of Zeus. Popular mythology has made us too familiar with the notion of the comic-book Olympian hurling thunderbolts at all and sundry for us not to be taken by surprise by a description of the original nature of his weapon:

> [a] magical instrument of domination; through it Zeus can "tame" his divine enemies. . . . Hesiod and other poets who followed him describe the terrifying effects of this shaft of fire. . . . They use images of cosmic disorder. The air catches fire, the waves and the ocean seethe, the earth, sea and sky collapse into each other. Tartarus ["the Underworld"], undermined in turn, shudders. All the different regions of the cosmos and all the elements are once again mixed up in a confusion resembling the primordial chaos. The power of the thunderbolt is such that it reduces the world to its so to speak "original" state and therefore the victory it brings Zeus symbolizes a complete reorganizing of the universe.[7]

The classic image of this object, which appears repeatedly around the world as an instrument of cosmic destruction and regeneration, is the three-pronged "trident," often doubled-ended. It is shown in many Assyrian carvings that depict the confrontation between the sovereign and the winged monster and is commonly regarded as representing a "thunderbolt." In Buddhist iconography, the thunderbolt form is the *vajra,* "signifying the spiritual power of Buddhahood [indestructible enlightenment] which shatters the illusory realities of the world."[8]

It is just such a weapon that Indra employs against the dragon Vritra: "I will tell the heroic deeds of Indra, those which the Wielder of the Thunderbolt first accomplished. He slew the dragon and released the waters."*

---

*Wendy O'Flaherty, *Hindu Myths,* 74. A later verse says, "At that very moment you brought forth the sun, heaven and dawn."

Fig. 28. The Weapon of Shiva, trisula, *the "three-jointed weapon" that he fixed "upon the sky" for the destruction of the Asuras, whereupon "the mountains tottered, the earth shook, the depths of the sea were destroyed."*

Fig. 29. King destroying the Chaos Monster. Nineteenth-century drawing of an alabaster panel in the British Museum, from the palace of Ashur-nasir-apal II (885–860 BCE), Assyria.

To this colorful and varied armory, we could add the Minoan double ax, the hammer of Thor, Indra's horse's head, and a number of other unlikely weapons. There is hope of clarification, however. Santillana points out that all these weapons are star groups, although there are a number of opinions as to which is which. Manilius describes Zeus's bolt-bearer in his star-poem: "Towering Eagle next doth boldly soar as if the thunder in his claws he bore. He's worthy Jove since he, bird, supplies the heavens with sacred bolts, and arms the skies."[9]

The eagle described here is the constellation of Aquila, whose stars are the Hindu *nakshatra,* sravana, pictured as a trident. (The twenty-eight *nakshatras* are the Hindu star-groups into which the path of the moon is divided.) According to Richard Hinckley Allen in his *Star Names: Their Lore and Meaning,* "[T]he weapon in the hand of Marduk" is made up of the stars of the Scorpion's sting, which are also associated with the "smiting of lightning."[10] On the other hand, in India the

Horse's Head consists of the two brightest stars of Aries, the nakshatra known as Mg, and these stars are also the Gadagada, the Mace of Vishnu, "Time the destroyer." Since these stars are familiar to us as part of the Ram of Aries, it is worth pointing to the images from the Fon people of West Africa of the deity Xevioso as a ram who vomits thunderbolts. These "thunderbolts" are pictured as axes, and this is a common substitution for the trident.

The essential act of this weapon is that of dividing: lightning, thunderbolts, Thor's hammer, the Minoan double ax, and many others all represent the means of dividing chaos. That there are so many opinions about the precise location of these starry weapons should not be surprising. When the myths talk of the weapons with which chaos is defeated, they are describing the stars that act as the new determining elements of the Frame of Time. There is thus a succession of these determining star groups as the precession moves along its path, and each one bears the same title. The stars of Aries make up the ram that heralds the age of Zeus and of Moses. But the title of Prince, *i-Ku* in the Mesopotamian languages, has been applied to all the stars in succession from the Milky Way through Taurus to Pisces, and often to their "accomplices" on the opposite side of the sky as well.

Thus, Allen gives the title "Ku meaning prince or leader" to Orion, Capella, Aldebaran, and Aries. In his opinion, the word means "foundation, leader," but is also used for animals such as dog, cow, snake. The stars, he suggests, are seen as sheep (or sometimes camels). As their leader, Ku, the star that begins the year and determines the World Age, became *the* sheep and ultimately, as the leader of sheep, the Shepherd.

But the definitive and, so to speak, original thunderbolt/ax/mace remains the stars of the Hyades that make up the Bull's head. Here the world was first reestablished after the demise of the Golden Age, and it has marked the beginning of time ever since. In the extraordinary shrines that form such a major feature of one of the world's oldest towns, uncovered at Çatalhüyük in Anatolia, were found quantities of altars in the form of bulls' horns. Each has a socket for something to be inserted, perhaps a symbolic tree, a pillar, or a double ax. With this dividing mark of time in place, the altar would take on its sacred character. Something similar seems to be represented by the Egyptian hieroglyphic of solstice–new year, in which a spread-horn figure, representing the year, is divided by a pole with ribbons attached, representing a "god."

So profound was this relationship between the Bull and the Great Year that Hindu Dharma, "the divine law," was described as a bull. Each of its four legs represents an age of the total cycle. In the Golden Age it stands on all four feet. During the course of the four ages, true spirituality becomes increasingly obscured until the cycle closes with a catastrophe, after which the primordial spirituality is restored and the cycle begins once again. Not only is this a remarkable image, but it is clearly one with immense robustness as well as it turns up again in North America, where Sioux tradition tells that "at the beginning of the cycle a buffalo was placed in the west to hold back the waters. Every year this buffalo loses one hair and every age he loses one leg. When all his hair and all four legs are gone then the waters rush in once again, and the cycle comes to an end."*

---

*According to the inscriptions in the tomb of Seti I, the Egyptians similarly saw the Sky, *Nut,* as a cow; the "props of heaven" are established to hold her up (Wallis Budge, *Legends of the Gods,* at www.sacredtexts.com/egy/leg/leg05.htm).

HEAVEN ON EARTH

DEO GRATIAS

In the Irish stories that tell of the battle between the Tuatha De Danann (the peoples of the Goddess Danaan) and the Fomoroi (the demons from the sea), the gods have been hampered by the wound inflicted on their king, Nuada. To defeat the demon leader Balor, "lord of the underworld," and his armies, the Tuatha call upon the services of Lug, the master of all the arts, whose birth has predestined him for this role. The story is closely related to that which tells how at a crucial point in the battle between the Hindu gods and demons, the gods, who are losing the battle, decide that they need a "raja," a king, to help them. They choose Indra for the task.

From this kind of imagery is derived the model for kingship everywhere: the king's purpose is the maintenance of order in the kingdom and the suppression of chaos. As such, the king's role was directly related to that of the ruler of heaven; all the aspects of his life were determined by this need to imitate the celestial court and the processes displayed there. Indeed, it was from heaven that the king received his authority. Thus, the king in Chou times in China was called the Son of Heaven because, provided he remained virtuous, heaven bestowed on him its mandate to rule on earth.

Even in much later times, this notion prevailed, and the king considered his role in the state as directly related to that of the ruler of heaven. According to Margaret Murray, James I of England described his role in these terms: "For my part, I hold it the office of a king, as sitting upon the throne of God, so to imitate the *primum mobile* and by his steady and ever constant course to govern the changeable and uncertain motion of the inferior planets,"[1] and still today the monarch of the United Kingdom rules "by the grace of God."

To be granted this mandate, however, does not ensure a perpetual right to it; a king could, by his own misdeeds, disqualify himself and find his farr had left him, and with it his right to rule. For the Chinese emperor Yu, who had ordered the kingdom according to the Great Plan he had received, all was well for as long as he maintained what the text calls his "virtue." Should that virtue decay, however, and it inevitably must, then the numbers of the calendar would come in their incorrect order, and disaster would ensue for the state. Something similar happened to the Persian Jamshid, as described in the *Shahnameh*. During a reign of plenty, he had "organized the people into four castes, the priests, the warriors, the artisans and the cultivators . . . and so years went by until the royal *farr* was wrested from him. The reason for it was that the king, who had always paid homage to God, now became filled with vanity and turned away from Him . . . (saying) I recognise no lord but myself. It was I who adorned the world with beauty and it is by my will that the earth has become what it now is. . . . And now that you are aware that all this was accomplished by me, it is your duty to entitle me Creator of the World."[2]

"But," says the *Shahnameh*, "as soon as he had made this speech the farr departed from him and the world became full of discord. God having withdrawn his favour from Jamsyd . . . strife and turmoil erupted and glorious bright day turned to darkness. . . . Jamsyd's destiny was overcast with gloom and his world-illumining splendour disappeared."[3] The result is that he is sawn in two by the monster Zahhak.*

Any sign of the king's inability to govern according to the laws of

---

*In fact, the *Shahnameh* says that Zahhak had himself been corrupted by the devil, one of whose ploys was to teach him to eat meat. When it left Jamshid, the farr is said to have "fled to bottom of the ocean."

heaven was to be regarded as a sign that divine favor had been withdrawn, and, with it, his right to rule. An ancient Irish tradition tells of a silver cup that was kept by a public fountain; its continued presence was a sign that the king still retained the authority to rule. Irish princes were instructed that it is "by the prince's truth great peoples are ruled. For it is the prince's falsehood that brings perverse weather upon the wicked and dries up the fruits of the earth,"[4] and with what we have learned of the tilting of the cosmos, we can appreciate the awful implications of such "falsehood": "The part of the house of Tara [a palace that itself had been designed to represent the cosmos] where a king pronounced a false judgement slid down the deep declivity . . . and so will abide forever."[5]

Time itself, however, seems to have been sufficient to remove the most traditional kings from their posts. Sir James Frazer's renowned work on sacred kingship explores the wide variety of factors that might explain the process of ritually removing a king from office after the decreed period of time. Frazer, however, was considering only what we might call "terrestrial" time; a relationship between the king's reign and the greater cycles of the heavens does not figure strongly in his evidence, but now seems the most reasonable explanation. The king's reign represented on a human timescale the reign of the lord of the age.

According to Eliade, "Among the Fijians the installation of the chief is called 'Creation of the world'":[6] "Almost everywhere a new reign has been regarded as the regeneration of the history of the people or even of universal history. . . . The Indian ceremony of the consecration of a king is only the terrestrial reproduction of the ancient consecration which Varuna, the first Sovereign, performed for his own benefit—as the *Brahmana* repeat again and again."[7]

There is evidence that this human timescale representation of the celestial cycle was regarded as beginning and ending with each New Year ceremony. According to a Sumerian hymn of the second millennium BCE, the marriage of the king with the goddess Innana, "the Lady of the Palace," was consummated "at the New Year, on the day of decisions," in the bedchamber of a chapel built for the goddess.

A union between the king and a goddess (whose prerogative it was to dispense kingship, as we have seen) was an essential part of the royal ritual in ancient civilizations of the Near East, either at the New Year festival or at the king's coronation, or, as often seems to have been the case, at a combination of the two.

The notion of the "marriage" of the king to his realm survived at least into medieval Ireland, where the inauguration feast of a king was called a wedding feast. In his work *Heaven's Mirror,* Graham Hancock tells of the king of the Khmer at Angkor who had congress with a dragon, suggesting that this refers to the king's observation of the stars of the constellation Draco. The stories told in Ireland of the union of the king with a horse refer perhaps to something similar; in this case, the horse would most likely be the horse the Hindus called "the Year," that is, the whole of the precessional cycle. Such an experience might be similar to that described by Marsham Adams as being available at the winter solstice from the roof of the temple at Denderah, where during one night the stars of the ecliptic would travel overhead, the ecliptic pole being then located on the northern horizon.[8]

The cosmic role of the king is described most fully in the rituals attached to that king in India known as the *kakra vartin*—"the turning wheel"—the embodiment of the eight-spoked wheel of Dharma, the law of order and balance. At his inauguration he represented the central pillar himself, in a ceremony of cosmic regeneration. During this ceremony an actual pillar was set up called the Yupa, "the sacrificial post," which is likened to the "eight sided thunderbolt, the intermediary between the divine world and earthly life." A wheel is attached to the top of this pillar and a ladder* placed against it, which the king climbs, saying, "Come let us ascend to the sky." When his head is above it he says, "We have become immortal."

From the sacrifice that accompanies the ceremony, the king is considered to be ritually reborn. He is then anointed with the waters of Varuna, the lord of Rita.

---

*This ladder to immortality was discussed in "The Ladder of the Gods," page 98.

The idea of the king representing the cosmic pillar reappears in a number of places, often in the form of having the king stand on one leg. At the Temple of Horus, at Edfu in Egypt, there is a carving of the pharaoh standing on one leg at the ceremonial inauguration held on his thirty-year jubilee. Similarly, according to Santillana, the temporary king of Siam, set up for the yearly expiatory ceremonies, had to stand on one leg upon a golden dais during the coronation ceremonies. This "mock king" was known as "the Lord of the Celestial Armies."* Such performances remind us of Prince Dhruva, whom we met in the early part of this book, elevated to the polar axis as reward for his remarkable feat of standing on one leg for a month (see page 43).

RITUAL FERTILITY

On the Hill of Tara, which is situated in the center of the Plain of Fal (an ancient name for the whole of Ireland), stands the stone of Fal, later known as the "member of Fergus" or "stone penis," and the emblem of kingship in Ireland.[9] In India a similar stone was set up as "the essence of kingship." This is the *linga,* or "phallus," placed on top of a pyramid, in the very center of the royal residence. It was here that the axis mundi was supposed to reach the earth. Thus the stone penis reveals its cosmological origins, and so-called phallus worship must be reconsidered. Indeed, it is clearly necessary to reconsider the whole notion of what a "fertility ritual" might be concerned with.

In the opinion of Eliade, "The appearance and disappearance of vegetation were always felt, in the perspective of magico-religious experience, to be a *sign* of the periodic creation of the universe."[10]

The transfer of this perspective toward one of "fertility," in the sense in which it is more commonly understood, is, again according to Eliade, to be seen as one of those tendencies "often to be found in the

---

*Sir James Frazer tells us that this Siamese king stood "leaning against a tree with his right foot resting against his left knee" (J. G. Frazer, *The Golden Bough,* London: Macmillan, 1941, 284). This is the position adopted by the pharaoh at Edfu.

history of religion . . . the tendency towards the concrete: the turning of the notion of 'creation' into that of 'fertility.'"[11]

If this is so, it runs quite contrary to the popular opinion that notions such as fertility are the origin of religious practice, and that New Year rituals are concerned with rejuvenating the year, or with rescuing "summer" from "winter."* We are suddenly faced with a much more awesome meaning. The rejuvenation that is the concern of the myths and rituals is that of the world itself, as is repeatedly made clear in the New Year festivals. It is the fertility of the relationship between heaven and earth, and the resultant generation of a new creation, that is invoked.

This "fertility" was represented in India by the erection of the *lingam/yoni,* the symbol of the union between the stars, regarded as female and represented as circular, with the axis of earth, represented as a pillar penetrating the circle. Cosmic time and its measurement are the sole preoccupation of these ceremonies, in order that social life might be conducted in accordance with celestial law.

There is even the suggestion that human fertility was regarded as one further imitation of this celestial procreation, with the ceremonies of marriage closely following the pattern of the ceremonies that celebrated the re-creation.[†] This is born out by the use of the ceremonial marriage of Surya (the daughter of the sun) to Soma (the moon and the Draught of Immortality) as described in the Rig Veda Hymn to Surya[12] as the basis for all Vedic human marriage ceremonies.[‡]

Similarly, for the Ngayu Dayak people of North Borneo, the wedding ceremony is regarded as a reenactment of the creation and the reenactment of the emergence of the first human couple from the tree of life.[13] Each

---

*Eliade also emphasizes this notion: "The sufferings, death and resurrection of Tammuz, as they appear in the myths and in the things they reveal, are as far removed from the 'natural phenomenon' of summer and winter as are *Madame Bovary* or *Anna Karenina* from an adultery" (Mircea Eliade, *Patterns in Comparative Religion,* 246).

†For the significance of the patterns of choice of marriage partner, see page 188, "The World Divided."

‡This is the hymn that refers to the mystic third wheel of the sun, known only to the "inspired."

of these sanctified individual ceremonies is a small-scale reflection of the kind of annual performance described by Granet as "the ancient annual ritual of communal marriage. . . . Men and women grouped themselves into two rows facing each other and competed in stimulating antiphonies analogous to the way in which light and shade mutually complement each other. . . . The conclusion of the festivities constituted a communal marriage, a hierogomany for which the rainbow served as a symbol."[14]

The traditions of the Dayak people also have an extraordinary light to cast upon the whole business of New Year and fertility rituals. For them, the two or three months between the appearance of the stars of Orion (which the Dayak know as Patendo) and the beginning of the harvest constitute the sacred era of the world, the "Time between the years." Their ceremony, held as a harvest/New Year feast, is revealing enough to merit a substantial quote from the text in which Eliade provides a detailed description:

> It is not only that another harvest has been brought in, or that another
> year is passed; there is much more to it than this, for a whole era in
> the existence of the world has elapsed, a period of creation is ended,
> and the people return not only from their fields to their village but
> they return also to the primeval time of myth and the beginning of
> everything. . . . This is most clear in the lifting of all secular regula-
> tions and in the submission to the commandments of mythical antiq-
> uity . . . after the expiring of the world-era (the old year) the creation
> is re-enacted and the entire cosmos renovated. It is the time in which
> Jata [the deity of the underworld] emerges from the primeval waters
> and Mahatala [the deity of the upper world] descends from the prime-
> val mountain, when both are united, personally and in their totemic
> emblems, in the Tree of Life, from which the new creation originates.
> (The erection of a Tree of Life is one of the most important acts in
> the whole ceremony.) . . . The sexual exchange, total and mass, is not
> contrary to "hadat" [the law, custom, right behavior] . . . it is the union
> of Upper and Lower worlds . . . It takes place in accordance with the
> commandments of the total/ambivalent godhead.[15]

Such a description is shattering in its implications for our understanding of the possible original context of those traditions that have survived in less complete forms. As an example, we might consider the English May Day rites, which we are told once included sexual "license" as well as the erection of a Maypole, procedures presumed to be designed to increase the fertility of the earth but that now must be totally reconsidered.

THE POWER MAKING THE WORLD LIVE AGAIN

A vital part of these regenerative ceremonies held around the world during the "time between the years" was the reading of the "Decrees." Eliade describes the Babylonian New Year ceremony, the *akitu*, "The Power Making the World Live Again": "Within the akitu ceremonial the 'festival of the fates' was celebrated, in which the omens for each of the twelve months of the year were determined, which was equivalent to creating the twelve months to come."[16]

This in itself, we might suggest, was equivalent to creating the whole of the precessional cycle. At each New Year, the gods were considered to fix the destiny of the following twelve months; accordingly, the Sumerian hymns call the New Year "the day of decisions." The "determination of the decrees" was accomplished by the act of *nam-tar,* wherein the decisions taken were constituted and proclaimed. The Hebrew feast of the Tabernacles and the Persian Nawroz embody the same idea in various forms; and in their book *Celtic Heritage,* Alwyn and Brinley Rees describe the ceremony as it was enacted in medieval Ireland:

> Here too is the reason for which the feast of Tara was made at all; the body of the law which all Ireland enacted then, during the interval between that and their next convention at the year's end none might dare transgress. ... The festival was a return to the beginning of things. There in the presence of the kings, the nobles and the people, the whole mythological and chronological past of the nation was conjured into the present. This declamation clearly had the force of a creation

rite to re-establish the foundations of the tradition at the inception of a new period of time. The social order was also reaffirmed.[17]

A remnant of this notion survived among the European peasants until quite recently, whereby it was maintained that the weather for each month of the coming year could be determined from that of the twelve days of Christmas, especially the likelihood of rain.

It is clear that the New Year ceremonies imply the regeneration of time on a scale much greater than the explicit annual one. This point is emphasized by Eliade in his work *The Myth of the Eternal Return:* "A periodic regeneration of time presupposes a new creation, that is, a repetition of the cosmogonic act."[18]

Such large-scale ceremonies may be stretched over periods long enough for them to escape the attention of observers. Dominique Zahan explains that, in the African traditions,

[t]he world is subjected to the restorative action of man, who intends thereby to give it a new brilliance, a rebirth of life. Undertakings of this type are not well known to Africanists because of their cyclical character lasting considerable periods of time. However, there is no doubt that all the ceremonies for the renewal of the year, as well as those consecrated to the renewal of the kingdom among African royalty, once included and do to this day rites concerning the restoration of the cosmos. . . . The Dogon rite of remaking the world concerns the equinox and Sirius, takes 6 years, 3 in the upper region (north east) representing Anima the sky and three in the lower (south east) representing Nommo the earth and the whole ceremony is felt to be the restoration of the cosmos and of its cardinal points. . . . In his approach to the cosmos, the African behaves as an organizer of space and he tries to act on the universe, periodically remaking it, and thereby he believes bringing about its constant rejuvenation. . . . From what has survived . . . it can be inferred that the idea of cosmic renewal was profoundly anchored in traditional African religion.[19]

The whole purpose, then, of not only the ceremonies of New Year but also those of kingship and of "fertility" was the rejuvenation of the cosmos and the reestablishment of its proper order. But the intimate relationship between human life and the cosmos was more pervasive than the ritual feasts. It formed the basis for the whole pattern of organized life. Eliade identifies here what he refers to as a "primitive" ontological conception: "an object or an act becomes real only insofar as it imitates or repeats an archetype."[20] Nowhere is this conception more clear than in the "repetition" of the ultimate "archetype," the organized cosmos.

BUILDING THE KINGDOM:
AS ABOVE, SO BELOW

It needs must be that every nature which lies underneath,
should be co-ordered and full-filled by those that lie
above; for things below cannot of course give order to the
ordering above.

KORE KOSMOU, OR VIRGIN OF
THE WORLD; HERMETIC TEXT

The principle of "as above, so below" is the fundamental tenet of hermetic wisdom. This well-known dictum is usually taken to mean that
things will happen on earth as they do in heaven; the notion of astrology depends on it. In the mythical context, however, it seems this
dictum might be better understood as an injunction imposing an obligation on the inhabitants of earth to organize themselves as the heavens
are organized. The familiar words in the Christian Lord's Prayer, "Thy
will be done on earth as it is in heaven," express this obligation.

It is the duty of humanity to organize itself according to the heavenly law. This is the law encoded in the Egyptian concept of Ma'at and
the Hindu Dharma or Rita. Since the gods are responsible for the cosmic order, human beings must obey their commands, for these are based
in the norms—the decrees (the *me*) that ensure the functioning both of
the world and of human society. It is only by humanity conforming to

these cosmic norms that the world can proceed. As Eliade understands the traditional attitude: "The archaic world knows nothing of 'profane' activities: every act which has a definite meaning—hunting, fishing, agriculture, games, conflicts, sexuality—in some way participates in the sacred. . . . Every responsible act in pursuit of a definite end is, for the archaic world, a ritual."[1]

It is worth repeating here the remarks quoted earlier from Frobenius (see "Prime Thought," page 135); they succinctly describe the all-pervasive nature of this concept throughout traditional life: "For the whole of Black Africa we can now affirm the primordial importance of myth, both as epistemology and as the basis of family, social, political and even economic structures, which are nothing more nor less than exemplifications of [the] mythical."[2]

With the understanding we have established of the nature of "mythical patterns," especially the "twofold division of creation," we can look further into the way in which these patterns are exemplified.

THE HOUSE

The notion that human life is to be organized in accordance with the laws of heaven has been applied to the design of all aspects of civilization, from the definition of the nation, via town, village, and house design to the relationship between individuals and the community they live in. At the most fundamental and universal level, this can be seen in the design of houses. To demonstrate the ubiquity of the notion, we can take a quick tour of the ideal home from three continents. First, from the Uralian steppes comes this description of the nomadic dwelling: "The Samoyed Yuraks look on the central pole in their tents as an emblem of [the] column supporting the universe.* Certain ceremonies are performed in which the ascent to heavenly regions takes place *via*

---

*This column had to be kept in good repair; otherwise, the universe might collapse and the firmament fall and crush the earth. The story of why the Yuraks said the sky needed to be supported in the first place was told at the very beginning of this book.

this column and at such times the hole in the top of the tent represents the opening in the firmament through which the enchanter can reach the sky."[3]

From the tents of the far north to the plains of Nebraska is a substantial trip, but only a short step in design principles: "The earth lodge of the Skidi Pawnee was also a mirror of the cosmos. . . . [T]he Pawnee say that the floor is the earth itself and the ceiling is the sky. . . . The 'four pillars of heaven' literally hold up the sky-roof and the top of the sky . . . is the smoke hole."[4]

And finally from Africa:

> The single room of the first house [of the Fali people of Sudan] represents the primordial egg from whence issued man's earth, whose square form is reflected in the rectangular courtyard. The roundness of the room itself suggests the equilibrium of the nascent but already organized world, which contains its own future, and whose completion will be marked by the large oval family enclosure divided into four sections connected two by two. This is the image of a "finished world." The cycle will be completed by the circular shelter away from the family residence where the patriarch settles, detached from human activities and marking a return to the starting point.[5]

Given this relationship between house and sky, it is not surprising to find that a collection of houses represents an entire creation. Such a notion probably exists in all traditional settlements: two widely separated examples serve to demonstrate aspects of this notion. In his work *Vital Souls,* the anthropologist Jon Crocker describes the villages of the Bororo people of the Brazilian Matto Grosso: "The village is bisected by an East-West line paralleling the path of the sun . . . the moieties are to north and south of this line; each moiety has four groups each divided into 5–10 household groups." This village is seen as "the manifestation of transcendant reality," in whose eight divisions dwell the "totemic ancestors" of the clans. Such a village might well be regarded as being in opposition to the surrounding unoccupied land, as is well displayed in the African village: "The fundamental opposition seems to be that established between inhabited space and the uninhabited bush. The village is seen as a basic concept established in relation to the bush, the latter being devoid of order and open to chaos and confusion. It is thus that at the death of the family head or his principal wife the village is abandoned, since it is considered to be like the 'bush,' that is participating in primordial chaos."[6]*

The village thus represents the created world;[†] it has only a limited life span, in this case commensurate with that of the chief, but elsewhere of fixed time span: "A village is born, lives and dies. It is destroyed and reconstructed every 70 years. There is a world identical to the one of the living where the villages rebuild themselves. . . . The individual never dies, he simply rejoins the ancestors and his district."[7]

The logical conclusion is that this "mortal" village should represent an entire heavenly cycle, and so it does for the Tsuana people of Botswana, who gradually expand their village by adding huts around the circumference of a circle; once the circle is complete, the whole village is abandoned.

THE TEMPLE

Of all the built representations of the cosmos, the most predictable are those in which ceremonies and worship are conducted. The elaborately cosmic structures of the pyramid-builders in Egypt and in Central America are familiar, as are, perhaps, the temple structures built

---

*The "longhouse" of the Amazonian peoples is similarly abandoned upon the death of one of the grandfathers of the household group (see Reichel-Dolmatoff, *The Forest Within*).

†For the Fali people of North Cameroon, the villages in each district—or, more strictly speaking, the four hamlets that make up each village—are set out in a space divided so as to reflect the original fourfold division of creation. As the world is endowed with two contrasting movements produced by the rotations of the two eggs from which it arose, that of the toad and that of the tortoise, so one clan's dwellings (the Toad clan) are put round the opposite way to the others (the Tortoise clan); see Annie Lebeuf, "Le système classificatoire des Fali," in M. Fortes and G. Dieterlen, eds., *African Systems of Thought* (London: Oxford University Press, 1965).

by the civilizations of the Near East, which act as "communication points" between heaven and earth. The act of establishing such temples was in fact a reenactment of the original creation. At the temple of Horus at Edfu in Egypt, there is a depiction of the pharaoh and the goddess Seshat carrying out the ritual of "stretching the cord." The inscription reads: "I have grasped the stake. . . . I take the measuring cord in the company of Seshat. I consider the progressive movements of the stars. My eye is fixed upon the Bull's Thigh [Ursa Major]. I count off time . . . and establish the corners of the Temple."[8]

Speaking of the sacred nature of the temple's architecture, Krupp says, "The temple's environment . . . restated and emphasized cosmic order. The ceremonies performed there marked the pivots in the passage of cyclic time, those moments of re-entry into myth and encounters with the sacred."[9]

This reentry into sacred time was made possible because the temple itself, whatever its architectural conception, was regarded as the center of the universe, the place where creation was founded. Such is the tradition regarding the *eben shetiyyah,* the foundation stone that rested on the abyss at the center of the temple at Jerusalem. David, when building the temple, attempted to lift this stone and was told by a voice that it rested "on the Abyss"; removing it would bring a new inundation.[10]

But it is not only the grand temples of high civilizations that reflect this attitude. Krupp describes the temples built by the shamans of the Kogi people of northern Colombia: "The Kogi imagine their temples to be patterned after the spindle-shaped universe . . . they inhabit. Although they build only the conical, above ground section they think of their temple as continuing in mirrored segments towards the zenith and the nadir. . . . These are counterparts of the celestial and underworld realms, and the floor they actually build corresponds, quite naturally, to the earth. . . . On the temple floor are four fireplaces also positioned as the corners of a square at the intercardinal points."[11]

An opening at the top allows sunlight to penetrate. At the June solstice sunrise, it strikes the southwest fireplace, and at sunset that in the southeast. Throughout the next six months, the path it travels shifts northward until, at the December solstice, the light travels from the northwest fireplace to that at the northeast. The Kogi refer to this process as "weaving." The sun on its nightly "underground" journey weaves a black thread into the fabric between the white daytime ones.

This extraordinarily beautiful image of the "cosmic loom," in such a remote area of population, might well cause a reconsideration of all those mythical figures who preside over the process of weaving. The Kogi themselves, Krupp notes, see weaving as a sacred act: "Whether the 'cloth' that is woven is a garment, an individual life, or the universe itself makes no difference. All partake of the cosmic order, and so weaving among the Kogi is not just a practical enterprise; it is a sacred ritual act."[12]

One last example of a specifically "sacred" building will serve to sum up most of these ideas. The Delaware of eastern North America rebuilt their "Big House" annually: its erection, in October when the harvest was in, represented a ritual re-creation of the world and marked the beginning of a new year. The Big House itself, according to Eliade, was a reconstruction of the universe: "Its floor the earth, its four walls the four quarters, its vault the sky-dome, atop which resides the Creator in his indefinable supremacy. There are 12 carved faces, 3 on posts on each wall, plus those on the central pole, being visible symbols of the supreme power. The White Path, the symbol of the transit of life, runs from the east door to the right down the north side past the second fire to the west door and doubling back on the south side around the eastern fire to its beginning. Dancers 'push something along,' meaning existence, with their rhythmic tread. Not only the passage of life, but the journey of the soul after death is symbolically figured in the ceremony."[13]*

---

*The "White Path" mentioned here is the Milky Way. As a "symbol of the transit of life," this path is discussed further below (see "A New Path," page 209). The Tukano people of the Colombian Amazon have a similar "dance path" in their longhouse.

THE CITY

In the empires of both ancient China and South America, it was the emperor himself who was the center of the order of civilization: "He was the pivot of the world," says Krupp. "Fixed like the pole of the sky, he steadied the world, and all its affairs revolved around him."[14] It was in his palace that "earth and sky met"; it was his responsibility to preserve the world's order, and his capital was designed to facilitate this task by locating him at the center of the universal order. This unfortunate reliance on the "fixed pole" might perhaps be seen as an explanation for why none of these ancient empires survives. The old imperial city of Beijing, however, remains still defiantly aligned along its axis oriented by the pole: "Thirty-three hundred feet of Imperial Road reach from the Gate of Heavenly Peace [*Tien an men*] to the Hall of Supreme Harmony."[15]

Such a concept of celestial town planning did not stop at building layout. Evidence of the manner in which city planning of this kind could spread into the lives of all a nation's people is amply demonstrated by the Inca city of Cuzco, "the Navel of the World." Krupp describes the social organization of this city: "The Inca called their empire Tahuantinsuyu, 'The Land of the Four Quarters.' Its four geographic divisions were reflected in the empire's social and political organization, for the next level below the *Sapa Inca* ('Supreme emperor') in the rigorous social hierarchy was occupied by four *apusi* or prefects. . . . This four part division extended right into the centre of Cuzco."[16]

Inca society as a whole was broken into three groups, the royal aristocracy, their servants, and the general populace. Aristotle describes a similar-sounding system current among the Athenians, who had distributed their four tribes in imitation of the seasons of the year. By dividing each of them into three "phratries," they had twelve subdivisions corresponding to the months. Each phratry consisted of thirty families, making one family for each degree of the circle.

Such social planning was not restricted to location. Each population group might have a well-defined role to play in the life of the city. Within the city of Cuzco, for instance, there was a large number of sacred sites, said to be one for each day, and the maintenance and ritual use of each of these sites was the responsibility of a specific group of people.*

It is a long journey from the Inca city of Cuzco to the medieval English city of Southampton, yet this city preserves the remains of what might be a similar attitude to the arrangement of design and population in England. The division of a city's people into "quarters" (the Chinese Quarter, the Latin Quarter, and so on) is familiar enough. At Southampton, however, records remain of the manner in which the maintenance of sections of the walls of the city were allocated to the various trade guilds. These guilds each claimed patronage of a specific Christian saint and celebrated its annual festival on its patron saint's day. Since these saints' days are a calendar cycle, there appears to be a relationship between the calendar and the towers and gates around the city wall. Southampton is one of the few English cities where it is possible to reconstruct significant parts of this calendar.

THE NATION

Although the Chinese emperors and architects who planned Beijing built their city divided fourfold, for them the formative number was five, for they counted the four intercardinal directions together with the center. In their book *Celtic Heritage,* Alwyn and Brinley Rees, having mentioned these Chinese traditions and those of India, which divide the nation into four provinces with a fifth at their center and the people into the "Five Kindreds," compare this to the arrangement of Ireland, with Leinster, Ulster, Connaught, and Munster, respectively, at east, north, west, and south and Meath at the center.

Although the province of Meath is regarded as the central fifth in the Irish texts, Rees and Rees explain that the Ireland described in the medieval texts appears to have two centers, one at Tara and the other at Uisnech. The symbolism of these two centers and the relationship between them offers a remarkable demonstration of the ways in which

---

*The social structure of the Inca is discussed in "Ancestral Stars," page 191.

the cosmologies we have been exploring can reappear, for at Tara, the bounded cosmos is represented by four within four within four, whereas at Uisnech the cosmos, together with its source in the primordial chaos, is represented by five within five within five. Rees and Rees suggest that it was at the feast of Tara, which was held at Halloween, when the gates to the Otherworld stood open, that the marriage of the king with his realm was celebrated: "On this night of mischief and confusion, the four provincial kings and their people sat four-square around the king of Ireland, symbolizing and asserting the cosmic structure of the state and of society while chaos reigned outside."[17]

The Great Assembly of Uisnech, on the other hand, which was held at Beltaine (May 1), was "a reunion of the people," at which disputes could be placed before the Druids for judgment. It was here that the first fire had been lit: "From that fire were kindled every chief fire and every chief hearth in Ireland."[18]

The "Pillar of Uisnech" was five-ridged, symbolizing the five provinces at the center. Rees and Rees point out the relationship between this pillar, "the navel of Ireland," and that of Agni, the Vedic sacrificial fire, which is both the navel of the earth and "a pillar at the parting of the ways, a column supporting the five kindreds."[19]

Rees and Rees go on to show how the duality of Uisnech and Tara is strikingly paralleled by dual ritual sites both in Rome and in India.[20] In the descriptions of these sites that are recorded in the Brahmana scriptures, there are two essential fires and one accessory fire. The first, to the west of the sacrificial site, is the original hearth from which every fire is lit. This fire, which must be kindled with the fire-drill, is circular and is said to be "for the worship of the consorts of the gods." Its equivalents are the circular hearth of Vesta in Rome and the hearth of Hestia on Olympus. The other fire, kindled to the east of the first, is square and symbolizes the celestial cosmic world of the gods. In Rome, it is represented by the *templa quadrata,* of which the original city or its central symbol was the prototype.*

---

*These two fires can also be compared with those in the Delaware Indian Big House.

Rees and Rees sum up this duality of Tara and Uisnech with a revealing comment: "It is Tara that has the greatest resemblance to the Heavenly Jerusalem, which stands four square and oriented—a walled city from which sinners, sorcerers and the legions of hell are excluded. It would appear that Tara originally symbolized the cosmos of the gods as opposed to the chaos of the demons. Uisnech was primeval unity, the principle in which all oppositions are resolved."[21]

The notion of a nation as circular, embracing the whole round of the universe as "primeval unity," is most succinctly described by Black Elk, the Oglala Sioux elder, when he describes his vision of "the hoop of the nation."[22] At its center he sees "the red flowering stick," which he says is a symbol of the cottonwood tree at the center of the universe; this tree he calls *Waga Chun,* and we are immediately reminded of the Mayan *Wakah Chan,* the "Raised-up-Sky" tree, first erected at the creation and said to represent the Milky Way.

THE GREAT TRIBE

THE WORLD DIVIDED

Alongside the duality of Ireland, allowing the coexistence of the world of order and the world of unity, Irish traditions reveal another fundamental division, that between north and south. Almost as soon as the country had been occupied by the Sons of Mil, it was divided between 'the two brothers Eremon and Eber, and from that time throughout Irish literature, the northern half of Ireland is known as Leth Cuinn, "the Half of Conn," and the southern half as Leth Moga, "the Half of Mug." Speaking of this division, Rees and Rees refer to the fact that

> the division of societies into two moieties symbolized as upper and lower, heaven and earth, male and female, summer and winter, north and south and so on is worldwide, and the dichotomy is expressed in a twofold division of countries, provinces, cities, villages, temples, kindreds and other phenomena. . . . The superiority of one moiety over the other is, however, formal and ritualistic rather than political. The two halves have distinct and complementary duties and privileges . . . but there is also an antagonism between them which manifests itself in mock-conflicts, team games and other contests. Such a dualism seems to underlie the division of Welsh *cantreds* into "above" and "below." . . . "Great and Little," "within and without" or

"upper and lower" are commonly used to differentiate between two neighbouring villages or parishes . . . and rivalry between the two halves is common.[1]

Such remnants of structure are mere tatters, however, compared to those surviving in other parts of the world. For the Fali people of Sudan, according to Bastide, "humanity is divided into two clans, those of the Tortoise and the Toad." This world arose originally from two eggs, one that of a tortoise and one a toad's, which turned in opposite directions while their insides went around the opposite way. From these two eggs stemmed the first division of the universe. These two clans represent "the wild earth" and "the human earth."

We have already seen how Fali houses reproduce the primordial eggs and project at ground level the mythical geography, first binary, then quaternary. We can now see that this kind of division spreads out into all aspects of human life: "All the rules of kinship and all the matrimonial alliances likewise repeat or echo the various moments of myth, with the result that human existence down to its slightest detail falls within a mythical system and the actions of men both at home and outside continue and maintain the order and progress of the world."[2]

An apparently fundamental aspect of this "mythical system" is the division of the social group into two subdivisions, between whom "ritual antagonism" is the norm. Anthropologists refer to these large-scale social divisions as *phratries,* or *moieties.* The usual case is for tribes to be divided into two sets of clans, which are represented by "totems" whose characteristics are opposites: "The beings which serve as the totems of the two phratries have contrary colours. Where one of the phratries is disposed to peace, the other is disposed to war (among the Osage for example); if one has water for its totem the other has earth."[3]

The mythical events that result in these divisions are well displayed by the traditions of the Australian tribes of the Murray River. Nuralie is the name given by these tribes to a "group of mythical beings whom tradition places at the origin of things. The world was created by the

Nuralie; these beings, who had already long existed, had the forms of crows or eagle-hawks. The two species of Nuralie are represented as two hostile groups which were originally in a constant state of war. . . . There is also a sort of constitutional hostility between the totems of the phratries."[4]

In the light of what has been learned of the nature of creation myth by using astronomy as a guide, it is impossible not to consider this almost universal division of tribes and peoples into ritually opposed groups as being related to the division of the celestial beings whom the Hindu scriptures call the gods and the demons, and whose endless conflict occupies so much of the vast corpus of Indian mythology. Even without the widespread identification of these groups with their "totemic" animals, this would be a defensible position. Once the two groups are identified with the Eagle and the Crow, it seems irresistible, for these are the animal representations of Jupiter and Saturn across the world.

The ritual conflict enacted between tribal phratries, and mirrored in the structures of so many aspects of traditional life, is identical to that described in the Greek myths as occurring between Kronos and Zeus for mastery of the universe. When Zeus threw his father, Kronos, over the threshold of heaven, it was the inevitable conclusion of Kronos's attempts to resist the processes of time. From being the lord of the First Age, Kronos passed into the role of tyrant and eventually of outcast, although others said this was not so, that there was no battle, and that Kronos handed over power willingly, having grown old and weary and retired to the western caves where he continued to "dream what Zeus planned."

Whatever the preferred story, the implication is the same: Kronos, grown old, represents the past time, Zeus the new. Their conflict was the eternal one between order eroded into chaos and a new order imposing itself. The two halves of a divided tribe represent these notions, just as toad and tortoise represent "wild earth" and "human earth."

The successful propagation of the world depends on this rivalry, and this concept is reflected in the intermarriage and inheritance rules devised by many tribal societies. The resulting system has been well demonstrated by Anthony Jackson in his work *The Symbol Stones of Scotland.*[5] Jackson shows how inheritance might be passed from one phratry to the other and back in two generations, the intermediate generation's inheritance belonging to a man's sister's son. It is thus possible to see how the conflicts recounted in fairy tale and myth between uncle and nephew, or grandfather and grandchild, are the remains of this arrangement. In this way it is not only marriage that reenacts the original creative act; the interchange of generations is also made to play its part, and between them they ensure that the whole of human procreation both reflects and contributes to the patterns of regeneration displayed by the cosmos itself.

ANCESTRAL STARS

By means of this kind of regenerative interchange between the "rival" parties, members of these divided tribal systems are able to regard themselves as part of a unity that embraces the whole creation. Thus, according to Durkheim, "The Australian looks upon the universe as the great tribe, to one of whose divisions he belongs; and all things animate and inanimate which belong to his class are parts of the body corporate whereof he himself is a part. . . . Things are ideally distributed among the different regions of space, just as the clans are."[6]

That is to say, as Durkheim interpreted it, the Aboriginal Australians see themselves as part of a web of relationships linking all things in an ordered manner, and it is their totems that identify the particular nature of their places in that web. It is by means of their relationship to their totem that individuals know how to participate in the totality. In describing the totemism of the Native Americans, Durkheim says, "The universe, as totemism conceives it, is filled and animated by a certain number of forces which the imagination represents in forms taken from the animal or vegetable kingdom. . . . All things, animate and inanimate, [are] permeated by a common life (*wakan*): and this life could not be broken, but was continuous. . . .

The totem is the means by which an individual is put into relations with this source of energy; if the totem has any power, it is because it incarnates the *wakan*."[7]

In view of this notion of the totems as emanations of the primordial unity, it is not surprising to learn that the totems themselves are said to have been originally part of the body of creation. Speaking of the Arnhem Land tribes, Durkheim says, "It is related that in this tribe the totems were only the names given to the different parts of Baiame's body at first. So the clans were in a sense the fragments of the divine body."*

In the same way as the clans were thus separated out, according to Durkheim, the individual members appeared: "Mangar kunja kunja made men. Before his time there were no men, but only unformed masses of flesh, in which different members and even different individuals were not yet separated from one another. It was he who cut up this original matter and made real human beings out of it."[8]

It is by separating out of this unformed mass that individual souls are created and allocated to their part in the ordered structure. But these original "people" cannot be regarded as the first humans. They are better understood as the Ancestors, the original "totemic souls." These fundamental souls are "uncreated ones," beings who were "not derived from any others." Their existence is from "an epoch beyond which the imagination does not go—at the very beginning of time."[9]

They were the ones who first arranged creation: "At this time the totemic Ancestors travelled the world, creating as they went." A time came, however, when, wearying of the process of creation, the Ancestors went into the ground, all at their own points, leaving behind as a sign of their departure a sacred well. This return to the ancestral waters marked the end of the Dreamtime and the beginning of time. Thus Lloyd Warner says, "Here, too, in the well, lie the totemic ancestors who died at the beginning of time."[10]

---

*Emile Durkheim, *Elementary Forms of the Religious Life,* 294. Baiame's principal wife is said to be "the mother of all the totems."

Nevertheless, there has to be a direct link between these ancestral beings and the individual members of the tribe. The manner in which this is established is obscure. It seems that it is these totemic ancestors who are considered to be responsible for the fecundation of human mothers at the moment of conception. By identifying which particular ancestor was responsible, a child can be allocated to the correct totemic group. Durkheim suggests that it is in large part the locality where the mother believes she conceived that is determinative; "each totem has its locality, frequented by the ancestor." A link is then forged between the child and its *tjuringa,* the sacred object that represents both the totemic ancestor and the individual soul: "Each individual soul, represented by the *tjuringa* object, is a portion of the collective soul. This collective soul has its origin in one or two Ancestors from which all subsequent souls issue."[11]

The concept of the relationship among the individual, the landscape, and the ancestral creation is only one part of the immense tapestry of Australian tradition. The deeper we look into this mythical structure, the more profoundly impressed we become with the complexity of it, until we are obliged to agree with Strehlow that it represents, indeed, "the vision of a mental construction more marvelous and intricate than anything on earth, a construction to make man's material achievements seem like so much dross."[12]

The Australian tribes, however, were far from unique in their concept of the relationship among clan structure, the individual soul, and the nature of creation. William Sullivan has shown that a similar system existed in South America relating lineage or community groups (*ayllu*) to entities known as *wakas,* which he relates to the stellar groups that are said to be creators of animal types and social groups: "Just as each *ayllu* descended from a star, the people of each *ayllu* would live in harmony with all others, in the same manner that each star or constellation lived in fixed harmony with all the other stars."[13]

Sullivan goes on to explain that these wakas were created at the same time as the stars and were sent underground by the creator to reappear at their appropriate geographical location. The significance of the close similarities between the Andean system that Sullivan describes, whose links with the stars are clearly defined, and that of the Australian tribes has yet to be explored. These two cultures, however, are hardly the only ones that envisaged this kind of relationship between social structures and the stars. A striking light can be shone on this relationship by remarks reported from the Fon people of West Africa, who know of a god named Fa: "a god with 16 eyes. Diviners know how to open these eyes to reveal the correct one of the doors to the future."[14]

By casting a collection of shells (usually sixteen) onto a board and reading the result, a diviner can deduce the appropriate verses to recite to achieve the desired result. Marie-Louise von Franz quotes a prayer to Fa that invokes "the birds and wild animals which march past in the night and speak in man's language," and these are the ones who "receive the number of those who will die and give the number of those who will be born. . . . It is by means of number that these 'doctor animals' of Fa possess their power over the living. By knowing his Fa a man discerns Mawu's will for him, revealed in a saying which Mawu composes for each birth. From this song the man learns which of the lesser gods Mawu wishes him to worship."[15]

It is difficult in this context not to see the "zodiacal" constellations in these prophetic "birds and wild animals,"* all the more so when we consider that it is widely regarded to be the case that the original creating ancestors are stars, as Andrew Lang recorded: "The progenitors of the existing tribes, whether birds or beasts, were set in the sky and made to shine as stars."[16]

Egyptians, Australians, Inuits, and South African Bushmen all knew that their ancestors were stars. According to Andrew Lang, in

---

*Something similar seems to be implied in the *New Sayings of Jesus* (Grenfell et al., quoted by Jung, *Psychology and Alchemy,* 323): "Jesus saith (Ye ask? who are those) that draw us (to the kingdom, if) the kingdom is in heaven? . . . The fowls of the air, and all the beasts that are under the earth or upon the earth, and the fishes of the sea."

the Australian story of the flood, which occurred when men forgot the teachings of their first initiation and became corrupt, "of the good men and women after the deluge, Prudgel made stars."[17]

Similarly, Sullivan, in an appendix to his *The Secrets of the Incas,* says of the Puelche of Patagonia, "It was held that there was a time when 'stars were people,' while these same people 'today are animals.'"* The Skidi Pawnee of the Great Plains knew the same, as H. P. Alexander reported: "Our people were made by the stars. When the time comes for all things to end our people will turn into small stars and will fly to the South Star where they belong."[18]

These Australian, Andean, and African myths seem to be describing an extremely sophisticated concept by which human beings are made aware of the aspects of creation that relate to their particular position within it. The awareness of such a place gives the individual a sense of meaning within the community: each individual has a role to play in the whole process of creation and can identify "as brothers" all the elements of creation that share that role. Because all creation is defined and organized by the cycles of the stars, the ultimate source of these relationships is the "Ancestral Stars" themselves. Such a concept seems to have survived, although barely recognizable, in our notion of astrological "birth signs."

As a result of their relationship to the cycles of heaven, not only do individuals have a place in the earthly creation; but they also have a path to the eternal, for at the end of its life, as Lloyd Warner describes, the Australian's soul must return to its ancestral origins in the totemic well: "The totemic spirit goes into the sacred water hole through the ordinary water at the top of the well into the subterranean depths and finally into the totemic waters beneath, where the Wongar ancestors live, becoming part of the sacred configuration."[19]

---

*William Sullivan, *The Secret of the Incas,* 366. Sullivan is here discussing the suggestion that stars were first named, in Paleolithic cultures, in relation to the seasonal behavior of animals. It was with the arrival of Mesolithic cultures, he suggests, that ideas associating people with stars began to appear.

THE ORIGINAL SIN

Although the Dreamtime is generally regarded as the sacred time of Australian spiritual life, both as the original time and as eternally present during the ceremonies, both Durkheim and Strehlow tell stories that, in different ways, speak of a time before this. Some of these stories tell the familiar story of a tribal deity who had once lived on earth for some time but withdrew to the sky when men began to ignore the teachings of their first initiators. He is sometimes presented in the form of a worn-out old man.[20]

These stories seem to belong to a tradition separate from those that tell of the Dreamtime. Mircea Eliade, however, relates a tradition of the western Aranda that the Dreamtime heroes themselves had formerly had associations with the original sky-world by way of a mountain. For some reason, the sky-god caused this mountain to sink, and so the Dreamtime heroes had to remain on earth.[21]

But the most revealing traditions are those such as that of the Kaitish tribe of central Australia. As retold by Mircea Eliade, these stories describe a First Being named Atnatu: "Because certain of his sons neglected his sacred services he expelled them from the sky. These came down to earth and became the ancestors of men."*

This is not the only Australian tradition that tells of the origin of the Dreamtime ancestors in the sky, but it is the only reported one that tells of their transgressions, a motif that should be familiar. Because cultural contact between Australian Aborigines and the people of West Africa is not generally acknowledged, it is astonishing to discover the same story told by the Yoruba. This story was recounted by John Wyndham in his 1921 book *Myths of Ifè*. "Before the world was made," we learn, "there reigned Arámfè

---

*Eliade, *Australian Religion,* 29. Eliade, quoting Spencer and Gillan, *The Northern Tribes of Central Australia* (London: Macmillan, 1904, 498 ff), says the authors describe "Atnatu, the primordial being of the Kaitish tribe . . . as preceding the *alcheringa* ['Dreamtime'] times, that is the epoch of the 'creation' . . . [Atnatu] arose up in the sky in the very far back past." Eliade adds that this "probably represents the historical sequence . . . in all the central, northern and north-western regions."

in the realm of Heaven amidst his sons." Arámfè sends away his sons, telling them to "[c]all a despairing land to smiling life [a]bove the jealous sea, and found sure homesteads [f]or a new race whose destiny is not [t]he eternal life of Gods." And so they set out: "Here the Beginning was from Arámfè's vales through the desert regions the exiled Gods approached the edge of Heaven." As they stand on the brink of the chaos below them, one of them asks, "Orísha, what did we? And what fault was ours? Outcasts to-day; to-morrow we must seek our destiny in dungeons."*

For the fullest details of this story, particularly of the nature of the misdemeanor of the gods, we must turn to the *Kore Kosmou* (*The Virgin of the World,* the oldest of the hermetic texts). While acknowledging that most of the "hermetic" writings date from the first to the third centuries, Brian Brown, in the introduction to his translation, suggests on the basis of internal textual evidence that this text cannot be later than the fourth century BCE, perhaps earlier. Here it is stated that Thoth, the Great God, made the original souls in heaven; their purpose was to move the degrees, each in their own station. But these souls grew proud and began to move from their stations, and so Thoth bound them to a life on the earth below; in response to their wails of distress, he agrees that they may redeem themselves through successive rebirths.†

This story of the transgressions of the sons of god and their exile,

---

*See www.sacred-texts.com/afr/ife/index.htm. In a footnote Wyndham adds, "According to the legends told in Ífè, the gods were not sent away as a punishment; but there is some story of wrong-doing mentioned at Ówu in the Jébu country." Ultimately, their work of creation completed, "from the sight of men the Great Gods passed," that is, they transformed themselves into various aspects of the landscape, all that is except Ógun, who, weary of wars and replaced by his warrior-son, leaves for the mists of the West: "Above grey clouds the Sun of yesterday Climbed up—to shine on a new order. . . . So passed Old Ógun from the land."

†Brian Brown, *The Wisdom of the Egyptians,* 1923 (at www.sacred-texts.com/egy/woe/index.htm). The same story is told in abbreviated form in William Smith, William Wayte, and G. E. Marindin, eds., *A Dictionary of Greek and Roman Antiquities* (1890). Concerning the Orphics, we are told, "Transmigration of souls was a cardinal feature of their doctrine. They believed in the original sin of man, *sprung as he was from the ashes of the Titans* [my italics], and that the human soul passed from one body to another—that is, from one charnel-house to another till the ingrained taint was washed out and the purified soul was translated to the stars."

told both in Africa and in Australia—cultures that, according to the accepted histories, cannot be connected—is perhaps the most astonishing of all the stories we have seen spread across the world. It seems that here we have reports of the "original sin" quite unlike the one contained in the Hebrew scriptures.

THE DIVINE PART

We have traveled a great distance along the road through the mythical landscape, using the light from the stars as a guide. We have seen that the notions of the creation, destruction, and redemption of the corrupted world have been modeled on the ancient awareness of the value of time as a source of order. We have also seen how human culture has sought to emulate this order in its own organization by re-creating at various levels of society both the principles and the details of the celestial processes. In this manner have both the inspiration and the observation of those earliest astronomer-poets found their way into our spiritual and social lives, to remain there today virtually unacknowledged. Throughout the quest described in this book, however, there have been allusions and suggestions that point toward a significance for this argument in the most personal and pertinent aspect of our individual spirituality. Of all the mysteries that the archaic view of existence has bequeathed us, perhaps the most elusive and at the same time persistent and entrancing has been the idea of soul. With this story of the original sin and the ensuing fate of incarnated souls, we finally reach the point where this mystery can be confronted. How are we to understand the meaning of soul?

"The idea of soul," says Durkheim, "seems to have been contemporaneous with humanity itself," by which I take him to mean that some such notion is already current in all the earliest records and in the thinking of all societies, even those that seem to have remained the least changed by time. These ancient ideas of soul do not necessarily match exactly the way the idea is understood today, however; translators and anthropologists have tended to use the word to describe various complex ideas. This is not the place to try to trace the origins and history of these ideas in detail. What we can do here is to identify a number of characteristics that all traditions agree on and try to describe how these came about. These characteristics can perhaps be summed up as the idea of the soul as one immortal part of an eternal whole, incorporated into a mortal body. The body itself has the task of preserving the soul and returning it to the eternal, intact and untainted by the influences that surround it and which threaten to lead it astray. The question of what and where the individual soul is when not incarnated has been historically a matter of some dispute. This is not the case, however, for traditional societies, who generally agree that there is a "spirit world" in which souls reside between incarnations.

Until very recently, the notion of soul has generally been considered more or less synonymous with that of spirit or of self, and few people could have agreed on precise and separate definitions for the three notions.* Every functioning human body, however, has the experience of a "self," and contemporary neurophysiology has come close to demonstrating the substrate of this universal experience. The existence of a "spirit world" is not so common an experience; it is the product of "altered states of consciousness" of one sort or another, whether this be dream or trance or meditation. The role that the "spectrum of conscious states," from "normal" to "trance," has played since Paleolithic times in suggesting the existence of this world has been well documented by David Lewis-Williams.† He argues that the neurophysiology that generates the visions that create the idea of "another world" inhabited by spirits is universal, as is the idea itself, although each culture has attached particular and localized descriptions and meanings to these experiences.

What Lewis-Williams has not done, however, is to explore what links there may be between this universal imagery and the equally universal concepts of cosmological history contained in the myths of

---

*Traditional societies frequently recognize two or three separate kinds of "soul"; they, however, have very clear ideas about their various natures.
†See particularly *The Mind in the Cave* and *Cosmos in Stone*.

the peoples he enlists in his argument. While we can have no certain notion of the stories Upper Paleolithic people told in their myths (if indeed they told such stories at all), all those people whom anthropologists describe as living on the "Upper Paleolithic horizon"—that is, living in a "stone-age," hunter-gatherer culture—tell stories of cosmic creation, flood, and destruction, stories that include the emergence of the ancestral souls.*

For the Australian Aboriginals, each individual soul, represented by the tjuringa object, issues from one, two, or more Ancestral souls who together form a collective whole.† For most central Australian traditions, according to Durkheim's informant, these fundamental souls are the "uncreated beings" who were not derived from any others but existed at a time "beyond which the imagination does not go . . . at the very beginning of time," a phrase whose meaning we now understand quite precisely.

In many parts of the world, these formative ancestors, as we have seen, were widely recognized to be stars. It is back to them that the soul must travel at the end of its time on earth, whether it be across the sky or through the water of the totemic well. This understanding is still common among the descendants of the Inca in Colombia. In *The Secret of the Incas,* William Sullivan recounts the story of how he was shown by a young boy, whose grandfather lay dying, the route that he expected the old man to take across the Milky Way on his way home.[22]

It is a notion that Plato would have instantly recognized: Having described how God constructed the universal soul in the form of two hoops intersecting at an oblique angle, he describes how the creator took the remaining "soul-stuff": "Then when he had made of them one mixture, he took and divided souls therefrom, as many as there are

---

*For one example among many, see the description of the creation myths of the Barasana people of the Colombian Amazon by Stephen Hugh-Jones, *The Palm and the Pleiades,* 267 ff.

†According to Eliade, the Australians recognize two types of soul: one, the "shadow" soul, leaves the body at death and goes into the bush; the other, the "totemic" soul, lives within the totemic image.

stars, and to each star he assigned a soul and caused each soul to go up into her star as into a chariot."*

Plato explains that the material for these souls was "not so pure as before." These souls are then "cast upon the instruments of Time . . . each one into that which was meet for it. . . . And he who would live well for his due span of time should journey back to the habitation of his consort star and there live a happy and congenial life."[23]

Before exploring the consequence of this "casting," it is worth returning to Plato's description of the process whereby the Same and the Different were joined (that is to say, the relationship between the equator and the ecliptic, heaven and earth). It had by no means been a natural process: "He took the three elements of the same, the other, and the essence, and mingled them into one form, compressing by force the reluctant and unsociable nature of the other into the same."[24]

Thus, according to Plato, were the "the Circuits of the Soul" bound within the "River of the Body," the same body that we have called the Frame of Time.† This is the essential incarnation for the cosmos, which now "cannot remain for ever without change." Once endowed with its less-than-perfect body, the creation, we might say, became mortal.

Many traditions tell in varying ways that it was at the moment of separation of heaven and earth that death came into the world. In Babylonian texts, death is part of the device that the gods prepare to avoid a repetition of the "overpopulation" that led to the original destruction

---

*Plato, *Timaeus*, 41. Plato's doctrine appears to have been shared by many of the peoples of Siberia: William Sullivan (*The Secret of the Incas*, 367) says, "The Goldi, the Dolgan and the Tungus say that before birth, the souls of children perch like little birds on branches of the Cosmic Tree." (Sullivan is quoting from unpublished seminar notes by Hertha von Dechend.) The same notion is clearly stated in Cicero's *Dream of Scipio*: "Men are likewise endowed with a soul, which is a portion of the eternal fires, which you call stars and constellations" (Tullius Cicero, *On the Republic*, Book VI, 15, at www.fordham.edu/halsall/ancient/cicero-republic6.html).

†Cf. Plato, *Statesman* [*Politikos*], 269: "Only the most divine things of all remain ever unchanged and the same, and body is not included in this class. Heaven and the universe, as we have termed them, although they have been endowed by the Creator with many glories, partake of a bodily nature, and therefore cannot be entirely free from perturbation."

of the First World. In Inuit tradition, stories tell how the women chose light and death rather than the dark timelessness of the First Age. Eliade relates an Australian tradition in which the Ntjikantja brothers, after pulling up the spear on which they had climbed into the sky, uttered a death curse upon the earth-dwellers; he sums up this idea by concluding, "Men had to die only because all connection had been severed between the sky and the earth."[25]

The mortal condition of humanity is thus to be regarded as a direct reflection of the condition of the whole of creation. That is to say, Man is a microcosmic copy of the whole created universe, which, because it was itself only an imperfect image of the eternal whole, would one day return to the boundless whence it came.

As Plato describes it, the link between heaven and humanity is more than this shared mortality. Each individual soul is made from a part of the universal soul. This is an opinion repeatedly expressed, both in Australian traditions and in classical theology: the individual soul, though within a mortal body, was immortal, since it was a fragment of the divine, universal soul, just as the "universal soul" was only partially enclosed in the perishable body of the world while the rest of it remained undivided outside of nature. According to Guthrie, this was a conclusion shared by the Pythagoreans: "The universe was one, eternal and divine. Men were many and divided and they were mortal. But the essential part of man, his soul, was not mortal, and it owed its immortality to this circumstance, that it was neither more nor less than a small fragment or spark of the divine and universal soul, cut off and imprisoned in a perishable body."[26]

We can now see the momentous importance of the kind of "cosmologically" determined tribal organization that we glimpsed in the previous chapter. The whole tribe represents the complete cycle of time; each individual, as a member of one of the clans, represents an aspect of that cycle—that is to say, one of the "ages of the world." Just as that age must come to an end, so must human life.

But the age is not considered simply to end. It returns to where it originated, the vast well of potential, the eternal boundless. Thus, the human soul can partake of what we have come to know as "immortality." Indeed, it might almost be said that it is the function of the soul to seek out this immortality, since to do so will ensure the redemption of the whole creation.

Because of the limitations placed upon the creator by the nature of the materials the creator had to work with ("the brute matter of Time," Plato calls it), the work of creation was one of compromise. And though God can periodically intervene to make good the consequences of his creation, the eternal harmony will not be restored until "the times have been fulfilled"—until, presumably, the perfect state of the cosmos returns with the completion of the precessional cycle.*

The manner in which God intervenes is variously described. In previous chapters, we have seen how he "takes hold of the rudder" of the erring world and puts things to order. We have also seen how, when the world has sunk to its lowest state, having been left to follow its own brute nature, God "sends his son" to put matters to rights, just as Prajapati takes a new form to rescue the world as it sinks beneath the waters. But the early alchemists preserved a further intriguing form of God's intervention: "But when the living father saw that the soul was tormented in the body, he sent his beloved son for the redemption of souls and set up a wheel with twelve buckets, which is turned by the rotation of the spheres and raises the souls of the dying."[27]

Clearly this wheel is the zodiac; the hope for the salvation of the original creation is the passage of the universal soul through all the ages, passing from one mortal "body" to the next, one by one through the determining constellations, the cycle of the zodiac stars. The hope of salvation for the human soul thus must involve a similar passage through the zodiacal mansions. The fate of creation and the fate of the individual soul are inseparably linked. The directive placed upon human existence is that it should strive to make that existence, and the creation in which it exists, as much like the eternal as is possible, thereby in some manner

---

*Valentinus says, "The final consummation will be achieved when all the spiritual seed dispersed among beings will have attained perfection."

"circumventing" the inevitable shortcomings of God's handiwork. Human existence is held to be both a consequence of and a cure for the awful compromise of creation.

This conjecture is at the root of all spiritual disciplines, from ritual initiation rites to Christian Communion. The human spirit is itself responsible for the redemption of the cosmic body and its reintegration into perfect harmony with the universal soul. But the human soul is subject to the same kind of experience as the cosmos; it must continually struggle against the forces of chaos, the "untamed" creatures who inhabit the badlands of the spirit, the lures of the flesh, and what Buddhist tradition calls "attachment," that is, the mistaking of "illusion" for the true "reality." This struggle is won through knowledge of the true nature of the "Real World."

To facilitate the acquisition of this knowledge, many traditions have constructed long and arduous rites of initiation. For the novitiates who participate in the Fulani (West Africa) rites, the path of knowledge is

> a progressive instruction in the structure of the elements of space and time, whose essence must penetrate the postulant. At the same time it is presented as a succession of tests symbolizing the struggle which he must undertake against himself with the help of God in order to progress. The postulant must successively enter twelve "clearings" which symbolize on one level the year and its twelve months and on another level, his movement over an area where in passing from one clearing to another he encounters the mythical personages who must instruct him. In addition he is put in contact with wild animals which symbolize the forces with or against which he must battle.[28]

It should go without saying that these twelve "clearings," as well as having these significances, are, most meaningfully, the twelve signs of the zodiac and thus the twelve ages of the world. The "wild animals" we have met before. The postulant is also said to be "put in touch with the principal plants which are encountered in the pastoral life." In many parts of Africa, these "principal plants" have decisive cosmological roles to play, especially as the "Eight Kinds of Grain" that arrived from heaven and that seem to be the determinants of the celestial divisions, whether spatial or temporal.

After the twelfth clearing, the initiate receives from the wife of Koumen (who himself personifies initiatory instruction) a small cord with twenty-eight knots—the stellar groups that mark out the lunar orbit—"whose succession must be consciously understood."

Such initiation rites—and we might well add to this description of the process in West Africa those Egyptian rites described in the Book of the Dead and others less clearly defined from around the world—are designed to prepare the soul of the neophyte for its ordeal after death, when it must confront the various guardians that stand along its path.* Without these rites, the future of the soul is grim indeed. Both Plato and Plutarch make this point clear: "the founders of the mysteries had a real meaning and were not mere triflers when they intimated in a figure long ago that he who passes unsanctified and uninitiated into the world below will live in a slough, but that he who arrives there after initiation and purification will dwell with the gods."[29]

It is apparent, from the African example quoted above—and those other sources that are available agree more or less explicitly—that these "ordeals" describe a journey through the stars, and that as far as it is possible to tell, these are the stars of the zodiac. The most detailed description of this journey is probably that contained in the Egyptian texts, which describe the Twelve Halls of the Duat (the "NetherWorld of the Soul"). It should be clear by this stage that a journey along the zodiac is a journey through time, and cosmic time at that. According to Zoroastrian tradition, "The paths of Time are the paths which the soul must traverse on its way from death to the judgment."[30]

---

*According to the hermetic text "A Great Treatise According to the Mystery," Jesus taught the desciples the "apologies," the passwords and signs by which they would have to make themselves known to each of the powers when they were ascending through the heavens (Doresse, *Secret Books of the Egyptian Gnostics*).

THE WAY HOME

HAPPY LAND

In the Rig Veda, Yama is the first of the immortals to choose a mortal destiny: "Who followed the course of the great rivers, and who discovered the way for many. . . . King Yama is the one who first discovered the way; this trodden path is not to be taken away from us; on that way that our forefathers travelled when they left us, on that way the later born follow each his trail."[1]

In the subsequent verses, the soul of the dead is exhorted to advance boldly "along the ancient path which our ancestors of yore trod, the path of Yama, that is, and you will see both King Yama and the god Varuna carousing on immortal food."

At the end of this path dwell the Ancestors. The Rig Veda and later texts refer to them as *pitris,* or "fathers." According to the Mahabharata, the pitris reside by a sacred lake that is the source of all rivers, with a sacrificial pillar in it. Those who dwell by this lake are said to be no longer bounded by the human timescale; nothing past or future is hidden from them. They are to be regarded "not as departed ancestors but as an order of semi-divine beings dwelling in their own celestial world who will be joined by all whose funerary rites have been properly performed."[2]

Such a future also lay in store for those Egyptians who had properly carried out the necessary rites. In chapter 189 of the Book of the Dead, the departed one claims: "I sit down along with the spiritual bodies who are perfect on the side of the lake of Osiris."*

In the Persian Zend Avesta, Yama has become Yima, the first man and the first king. The story of how he came to reside in the land of the dead will sound familiar. Ahura Mazda, seeing that the golden age could not last, warned Yima that a fatal winter was about to fall, and that he must carve out for himself a great enclosure (Vara) under the earth, into which he should bring "seeds of men, women and cattle, the greatest, best and finest," the most sweet-scented plants, and the most tasty and sweet-smelling foods. In the Avesta, this is the land of the dead, where all live happily with Yima. Later texts, however, say that Yima will remain together with his elect until signs foreshadowing the last days appear. Then King Yima and the noble race he has reared will emerge from under the ground to arrange the world again.[3]

This story is also told of Manu, who, like Yama, is a "son of the sun" (Manu Cakshusha, that is: there are to be fourteen Manus; we are born of the seventh, Vaivasvata). He had rescued and protected a fish that grew and grew until it "pervaded the whole ocean," at which point Manu recognized it to be the lord Vishnu himself. In return, Vishnu shows him a boat: "This boat," says the text, "has been fashioned out of the assemblage of all the gods, in order to protect the assemblage of great living souls."[4]

The god tells Manu that in a short time the earth will be submerged in water, and that when the ship is struck by the winds that blow at the end of the age, he is to use the serpent that will come to him in the form of a rope to fasten the ship to the horn of the god, who will appear as a fish: "At the end of the dissolution, you will be the Prajapati of the whole universe, moving and still. Thus at the beginning of the Krita Age you will be the omniscient, firm king, the overlord of the period of Manu, worshipped by the gods."[5]

In the Babylonian epic, when Gilgamesh goes to the underworld (he must cross the ocean to get there), he finds Utnapishtim, who has

*E. A. Wallis Budge, *The Book of the Dead,* 642; chapter 189. Faulkner's translation reads, "I will dwell with those potent noble dead in order to excavate the pool of Osiris."

a similar story of deluge survival to tell. Gilgamesh, however, has come plumbing the depths, although perhaps unwittingly, to claim the measures of a new age. According to Andrew George, "The wisdom he brought back from his journey was more than personal knowledge. . . . Evidently Gilgamesh was responsible for re-civilizing his country. . . . The Epic's opening words make the same connection: he 'saw the Deep (Abzu), the country's foundation.'"[6]

George makes the point that, although it is not said explicitly in the epic, "every Babylonian would know" that Gilgamesh was to become, after his death, the deified ruler and judge of the shades of the dead, the "lord of the underworld." It is to him that each soul must pass on death, to receive his judgment.

It is hardly likely to come as a surprise, however, to learn that for a soul, reaching the land of the dead is not so simple as the Rig Veda seems to suggest. This "First Way" that Yama opened lasted only as long as the First Age. In central Polynesia, it is Tiki, the first man, who opens a hole by which souls pass down to the other world: "It was by way of Tiki's hole that Maui descended into the home of Mauike in search of fire."[7] However, the passage he had used was shut from that day on. Thus, the souls had to find a new way "after his hole had been closed."

There are analogous myths in Madagascar. They form one of the variants of an almost universal belief in a visit made by the first men to the "other world," and consequently in the existence of a path between the earth and the world above that was ultimately destroyed by human wrongdoing.

It seems that what closes the barrier that prevents passage along an outmoded route is the ceasing of the "clashing rocks," which in their day had formed one of the hazards the hero must pass. These gates are a wonderful example of what Santillana calls "stone age boulders" carried along in the glacial drift of mythical stories. Hardly an obvious or natural notion, they reappear in the most diverse traditions, as we have seen. Anyone interested in tracing the path of this boulder-depositing glacier might add to the examples already mentioned: the clashing cliffs that guard the grounds of the Sagay ancestral shamans and "the place where the doctors go through" spoken of by the Australian Wiradjuri—"it kept opening and shutting very quickly." South American traditions replaced these rocks with a whirling wheel of razor-sharp knives; Brittany peasants spoke only of "the keyhole to hell," although the wren, intent on retrieving the lost fire, was burned while passing through it. They might also remind us of the "whirling sword" that Yahweh placed in Eden to protect the Tree of Life after the Fall.*

The Rig Veda calls this obstacle the "divine doors": "Thrown open be the Doors Divine, unfailing, that assist the rite, For sacrifice this day and now."[8]

Once the hero has passed through on his way to redeem creation, these gates do "fail." Some other route to the land of the dead must be found. Only the soul that knows this secret can hope to achieve its goal. The words of Christ immediately come to mind: "I am the Way. None comes to the Father unless it be by Me."

A NEW PATH

To trace in detail the routes to be traveled by the soul after death and the hazards met on the way would be beyond the scope of this work, consisting as they do of a bewildering quantity and variety of obstacles, rivers, lakes, seats, monsters, gates, and other diverting features. For our purposes, it will be sufficient to examine three principles. First, let us establish securely that this journey is conducted not simply "in the heavens"

---

*M. Detienne and J.-P. Vernant, *Cunning Intelligence in Greek Culture and Society*, 155, speaking of the Greek "Symplegades" through which Jason and the Argonauts must pass in search of the Golden Fleece, say, "They also move up and down, rising from the depths of the abyss of the sea right up to the sky. Situated at the furthermost confines of the world, these are the impassable gates, the columns to which form the pillars of the sky." Mircea Eliade describes various traditions of this hazard, including a "door standing at the spot where Sky and Earth meet" or where "the ends of the Year come together," both of which suggest that we might look for this door at the crossing of ecliptic and equator (*Patterns of Comparative Religion*, 247; *Australian Religion*, 132). Eliade adds in a footnote the account of another medicine man who "visited heaven through a small aperture revealed at intervals in the revolving two walls of a cleft."

but across the stars. Although by now it should hardly need stating, this notion is so alien to our Western concept that is worth repeating. We have already seen an example of this notion expressed in contemporary South America. It is similarly held to be so in almost every other tradition and has been so since the earliest records. Thus in the Book of the Dead, the departed proclaims, "I pass along the ways of those sky beings who determine destinies" ("the paths of the divine celestial judges").*

Christian tradition has whittled away at the notion of heaven until the very idea that it is to be found located in the sky has been dismissed as fantasy; other traditions remain more obstinate in their knowledge. Says Cicero: "But Scipio, cultivate justice and the sense of duty. . . . Such a life is a highway to the skies, to the fellowship of those who have completed their earthly lives and have been released from the body and now dwell in that place which you see yonder (it was the circle of dazzling brilliance which blazed among the stars), which you, using a term borrowed from the Greeks, call the Milky Way."[9]

The second principle that can be established is that, although there may be some confusion over whether the journey is made along the zodiac, along the Milky Way, or along the latter after the former, we can say with some certainty that there are widely recognized points of "crossing." Whether they are a direct connection between this world and the other or merely allow the soul to pass from the zodiac to the galaxy on its journey, these crossings can be precisely located. They are the two points where the Milky Way and the ecliptic cross, the "northern Gate" being between the constellations of Gemini and Taurus and the "southern Gate" between Sagittarius and Scorpio. To pass through these gates, say Santillana and von Dechend, they must stand "open," that is to say, they must rise above the horizon immediately before sunrise at the definitive times of the year.

According to Macrobius, "One is called the gate of men, the other that of the gods. Cancer is the gate of men, because through it they

---

*E. A. Wallis Budge, *The Book of the Dead,* chapter 43. Faulkner has, "My name is one who passes on the paths of those who are in charge of destinies."

descend to the lower regions; Capricornus is the gate of the gods because through it souls return to the seat of their own proper immortality and rejoin the company of the gods."[10]

Santillana and von Dechend explain that when Macrobius talks about Cancer and Capricorn, he is talking of zodiac *signs,* those of the solstices. They go on to specifically "locate" these signs at the intersection of the ecliptic and the Milky Way.

Such has been the opinion of most traditions. In the Andes the solstices are still seen as the time when the gates to the world of the dead stand open. There seems to be a general agreement that souls must pass through one or other of the "gates," "Capricorn" (i.e., Sagittarius/ Scorpio) for leaving and "Cancer" (i.e., Gemini/Taurus) for arriving, at the winter or summer solstices. A good deal of confusion surrounds this notion of the gates, much of it generated, as in Macrobius's case, by the use of the "astrological" system of division of the ecliptic, whereby "signs" rather than actual constellations determine the divisions. This system dates from the time when the stars of Aries marked the spring equinox, thus placing the solstices in the stars of Cancer and Capricorn. The "signs" were then locked into their positions in the calendar. Thus they became dissociated from the actual constellations rising with the sun throughout the year, as the precession moved the calendar forward.

The confusion, however, goes deeper than this, and the hazard is thus greater for any lost soul looking for guidance. As Santillana and von Dechend point out, "One would expect this station [the crossing from earth to heaven] to be at the crossroads of ecliptic and equatorial co-ordinates at the equinoxes. But evidently this was not the arrangement. A far older route was followed."[11]

By "a far older route," they mean older than the one they would anticipate. As we have seen, they were assuming a "beginning" dating from the era of the spring equinox in the galaxy between Gemini and Taurus, around 4300 BCE. The widespread tradition of using the solstices as gates seems to date from the era considered as a "beginning" by Hancock and Bauval, around 10,500 BCE. The concept of "beginning" described in this book, however, provides the "ideal" solution to this problem, since it locates the gates once again not only in the galaxy, but also at the equinoxes, as one might expect. The crossing from earth to heaven, after all, is most likely to be at the place where their respective circles intersect.*

However this matter is resolved, the crucial problem remains. The crossing between zodiac and galaxy, between this world and the next, is available only when the sun is in the galaxy, that part of the sky where the two cross, *at the appropriate season*. There are vast intervals of cosmic time when this ideal gate is not available. If a soul can only enter heaven when the crossings of equator and ecliptic (i.e., the equinoxes) are in the Milky Way, there are only two opportunities to cross in an entire precessional cycle, and the same is true of solstitial crossing points. If it is important which gate the soul uses (Macrobius's Cancer or Capricorn), then there is only one.

For the intervening times, other expedients are necessary. This is what the myths mean when they say that at the end of the golden age, the road to heaven was closed and the cycle of creations began. The relentless pressure of the precession pushed the equinoctial crossing out of the galaxy, and the direct route from earth to heaven was no longer available. A new route had to be found.

A soul leaving earth (and Plato's description implies that there were no souls before that time, since they were made at the same time that the universal soul was incarnated in the River of the Body; the myths certainly imply that such souls as there may have been were not subject to death) must take a route determined by the world age from which it came, for these souls are only mirrors of the universal soul, and they

---

*The notion of a "crossing" being available at the solstices seems to be derived from the model of the heavenly sphere divided into three by the tropics, the northern third representing the world of the gods, north of the tropic of Cancer; the central third representing the "inhabited earth" between the tropics; and the southern third representing the "underworld" south of the Tropic of Capricorn. When the sun reaches its highest point, which is at the June solstice, a way is available across to the land of the gods. This is held to be passable when the solstice coincides with the galaxy. This system is well described in William Sullivan's *The Secret of the Incas*. If the galaxy is regarded as a route from this crossing to the abode of the gods, then it leads to the zenith, not to the pole (see figs. 22 and 24). The whole model is thus based on a horizon/zenith system of "coordinates."

must "reincarnate" themselves, age after age, living out a further aspect of completeness, just as the universal soul itself must, coming to perfection at last only when "the times shall be fulfilled."

This is the reason why ancient texts say again and again that the "Happy Land" over which Yama rules, the "land of the dead," offers only a temporary lodging. A visit to the Ancestors is not forever. The Hindu text of the Prashna Upanishad explains: "The year too is Prajapati; it has a southern and a northern path. Those who revere sacrifice and meritorious works simply as acts secure only the lunar world for themselves; *they surely come back again.* . . . Those seers who wish offspring, betake themselves to the southern path. This assuredly is matter, the path of the ancestors. . . . The path to the *pitris* [the fathers] is the southern path, the *pitrayana,* the moon path."[12]

This land of the dead, then, is to be reached by a southern path, and various sources lead the authors of both *Hamlet's Mill* and *The Secret of the Incas* to conclude that it is located to the south of the Tropic of Capricorn, among those stars that the sun and the planets never visit. From here, "the souls see the South Pole," as Plato says, but it is not an eternal rest. As the Upanishad text says, "[t]hey surely come back," and this is the case for most traditions. After a long period with "the Ancestors," the soul must return to earthly existence.

There is, however, as the Hindu texts suggest, an alternative available to those for whom repeated incarnations are unwelcome: "The second path, the *Devayana,* leads to the gods. The sun is its entrance; after death ascetics and those who have attained knowledge and faith travel this path which leads to complete absorption in Brahma. It is in the north, beyond Mt. Kailasa, whence stretches a narrow, terrifying path barely wide enough for one person."*

---

*Margaret Stutely and James Stutely, *A Dictionary of Hinduism,* "devayana." The source is the Prasna Upanishad: "Those who seek after the Self by self-mortification, chastity, faith and wisdom secure the sun for themselves by the northern path. That is the abode of living creatures, which is the immortal, the final goal. Thence they never come back" (trans. R. C. Zaehner). The Vishnu Purana supplies more-precise locations of both this and the "southern path" (see Santillana and von Dechend, *Hamlet's Mill,* 407–8).

This path is not easily traveled, however, nor is it available to all. The earliest texts from Egypt, and some indigenous traditions around the world, imply that it is available only to kings and those of noble birth.* There is the suggestion here that it is only those who stand in direct line of descent from the gods, who represent them on earth, who can avail themselves of this option. Most traditions agree, however, that others can travel this path, providing they have prepared themselves. Having described how the Polynesian soul must depart at one or other solstice (depending on whether it comes from the north or the south of the island), Santillana and von Dechend describe the Polynesian belief that souls can stay permanently only when "they have reached a stage of unstained perfection, which is not likely to occur frequently."[13]

The Upanishadic text suggests that when this unlikely event does occur, the perfected soul can reach the gods via the "northern" route, passing over the narrow bridge that crosses the Tropic of Cancer into the northern sky. This northern path, which leads directly to the gods, and from which "they do not come back," can be attained by "self-mortification" and the attainment of wisdom, "the Knowledge of the Way." Pindar reported something similar: "Those who have persevered three times, on either side, to keep their souls free from all wrongdoing, follow Zeus' road to the end, to the tower of Cronus, where ocean breezes blow round the island of the blessed."[14]

The Gnostic texts call this "Tower" of Kronos the "Treasury of Light": "The Origin of Light judges souls, sending perfect ones towards the powers of the Treasury of Light. She sends the others back into the rotations of the celestial sphere or into the infernal abodes peopled with fantastic demons the long procession of which forms the body of the dragon of the outer darkness."[15]

Avoiding wrongdoing is clearly not sufficient to gain access to this

---

*The hermetic text *Kore Kosmou* says: "The king is the last of gods but first of men, divorced from his godship while on earth; his soul descends from a region above that of other souls. Those who have lived a blameless life and are about to be changed into gods, become kings that they may train for godship; or those souls who are already gods, but have slightly erred, are born as kings."

path; what equips a soul for a direct passage to the "Treasury of Light" is "wisdom." And the particular wisdom sought by the Gnostics was *gnosis hodua*—Knowledge of the Way. A soul possessed of this knowledge will have nothing to fear in the hall of judgment.

THE LAST JUDGMENT

At the climax of its passage through the divisions of the *duat*—"the Netherworld of the Soul"—having passed through the terrors of the first five divisions, the Egyptian soul is led into the "Hall of the Double Truth." Here the deceased must recite the forty-two "negative confessions" to pass the test of the "weighing of deeds." Then he must prove that he can call all the parts of the doorway through which he must pass by their proper names. Only then can he go forth to the "weighing of the soul in the balance," where his "heart" is measured against the feather of Ma'at. Graham Hancock has suggested that the image of the "heart" (it is in the form of a jar), by which the soul is pictured on the balance, represents "the Word of Thoth."[16] It was the god Thoth who gave "the Word which resulted in the creation of the World," according to Wallis Budge. He was the companion of Ma'at at the creation, and he alone could teach "not only words of power but the manner in which to utter them."[17] Thus the whole judgment process in Egypt was known as "the weighing of words," and it would seem that what was being measured was the supplicant soul's knowledge of these words.* Thoth himself stood by, ready to record the result of the measuring, the ultimate fate of the soul. Should this knowledge balance perfectly against the feather of truth, then the soul would be welcomed into the company of the Imperishable Stars. Of the perfected soul it is said, "He shall come

---

*E. A. Wallis Budge, *The Book of the Dead,* chapter 94: "Hail aged god . . . who art the guardian of the book of Thoth. . . . I am endowed with glory. . . . I am supplied with the books of Thoth." The text of the judgment scene contained in the Book of Gates says, "The examination of the words takes place, and he strikes down wickedness, he who has a just heart, he who bears the words in the scales, in the divine place of the examination of the mystery of mysteries of the spirits."

Fig. 30. The Hall of the Double Truth. Anubis reads the balance; Ammit waits to swallow the imperfect soul. On the right pan is the "heart" of the deceased, on the left the goddess Ma'at, and seated atop the pillar is the ape of Thoth. The texts are exhortations to Thoth and Ammit. Detail from the Papyrus of Ani.

forth in whatever form he is pleased to appear in as a living soul for ever and ever."[18]

Should the soul be found wanting, should the balance be less than perfect, then the monster Ammit, the "Eater of Souls," a mixture of crocodile, hippopotamus, and lion, sat crouching nearby to swallow it up into extinction. For the ancient Egyptians, this seems to have been final; the only alternative to becoming a perfected soul was *mut,* death. There appears to be no recognition of unperfected souls being "recycled" back into the created world.* Nevertheless, the soul-eating Ammit is reminiscent of the Gnostic dragon, the "Prince of this lower world" that prevented souls who were without Gnosis from reascending. He swallowed up the imperfect souls: "Passing through his body [they] were sent through his tail into the terrestrial universe where they were transferred into the souls of various animals."[19]

This was the tradition known almost everywhere across the world, with the apparent exception of Egypt. The "imperfect souls" that were without Gnosis were doomed to return to the "rotations of the celestial sphere," to be reborn as the souls of "various animals." These "animals" are to be understood in the same way as the various "wild animals" we have met before. They are, essentially, the constellations of the zodiac. Once committed to this process, these recycled souls must live through all the "rotations of the universe" until the times "shall be fulfilled." Assuming that they manage to live "decent lives" during this process, they will finally be perfected when the universal soul itself is perfected.

What, then, is the form of this knowledge, this *gnosis hodua,* that makes it possible to travel the path of the gods? It seems that what the soul needs to pass through the ordeal of the Last Judgment is a profound understanding of how the two gods Thoth and Ma'at, "the Word" and "the Plan," between them generated the world of becoming. The secret of eternal existence is the knowledge of how the time-bound world is generated from the eternal—knowledge, that is, of how the individual existence is joined to the universal: in short, knowledge of the precession and its resolution.

But it is not enough to have acquired this knowledge. For the deceased's soul to perfectly balance the Feather of Truth, it must not only

---

*In the New Kingdom texts known as "the Book of Gates" there appears to be a suggestion that "damned" souls could be reborn; at the Gate of the Sixth Hour, the serpent of chaos, Apep, is shown regurgitating those he had previously swallowed. Nevertheless, elsewhere in the same texts the damned are shown being tormented in the Lake of Fire. See www.egyptologyonline.com/book_of_gates.htm.

have knowledge of these matters, it must have experienced them, too. The "Trial of the Balance" is a test of whether the soul has "known" all the aspects of existence—that is, has been through all the stages of the full cycle of the precession, has lived the entire history of the creations. Only in this way could a soul become "perfected." For the Egyptians, this meant knowing the names of, and the answers to the questions asked by, the guardians of the various stations along the way. Something similar is referred to in the Gnostic text *According to the Mystery,* in which Jesus teaches the disciples the "apologies," "the passwords and signs by which they will have to make themselves known to each of the powers when they are ascending through the heavens."[20] The initiation rites we have described, with their persistent imagery of a passage through the stars, seem designed to enable initiates to live all the existences demanded of them during one life span, and thus become "perfected." As the Egyptian text puts it, "[h]e will go down to any sky."* The "self-mortification" that the Upanishadic text requires is the practice of "asceticism" of nonattachment, of keeping the mind set on the Real World and freeing it from "illusion."

AMBROSIA

One final process remains before the soul can become a god; it must be reunited with the body. To do so, it must be fed with the mysterious "celestial foods," the drug of immortality, whereby body and soul are reunited. In Egypt, the celestial foods (what the texts call the "*hu* and *tchefau* foods") were delivered by the gods Hu and Saa.† The Egyptian texts repeatedly refer to these foods as "cakes and ale." Something very similar survives in the Christian sacrament ritual. Speaking of the Holy

---

*From a spell contained in the ancient Egyptian Coffin Texts: quoted in Graham Hancock, *Heaven's Mirror,* 80.

†The "Hymn to Ra" contained in the Papyrus of Ani says, "Thou makest strong thy *ka* with *hu* and *tchefau* foods." Wallis Budge explains these foods as "the celestial foods upon which the gods and the beatified live"; cf. also *The Book of the Dead,* chapter 148: "Hail, ye who give cakes and ale and splendour to the souls."

Communion, St. Ambrose says that the transformed bread is *medicina,* the drug of immortality, "which in the act of communion, displays the characteristic effect in and on the believer . . . the effect of uniting the body with the soul."[21]

This is no longer the earthly body, however. It must undergo what Jung calls a *reformatio;* it is in fact a spiritual body, the building of which was a topic of special knowledge for the Gnostics.*

This healing and reformatio is the last essential process before the individual soul rejoins that of the universe. The Catholic Missal text says, "Grant that through the mystery of this water and wine we may have fellowship in the divine nature."[22]

For the Gnostics, the wine was not only the blood of God, but also "Grace." Marcus the Magician says that the wine is the blood of Moira, whose name means "degree" and is the Orphic name for "Intelligence." "Into all who drink of this wine," he says, "grace [*Charis*] will flow . . . and this Grace is Silence, the Mother of All, who together with the Ineffable Depth brought forth all the emanations of divine beings arranged in harmonious pairs."[23]

We might suspect that in this instance, if the "wine" represents the Silence (see "One and All," page 106), then the bread, "the Body of God," must represent the "Ineffable Depth." In the act of consuming the "celestial foods," the drug of immortality, supplicants appear to be feeding on both the "unity of the many" and the "unity of the One." In this way a soul can come to perfect knowledge, not just of the cycles of existence but also of the reality of eternity from which these existences are brought forth and into which they return. Such a soul can achieve immediate access to "the abode of the living creatures, the immortal, final goal": "He shall have an existence among the living ones and he shall never perish. . . . He shall not die a second time. . . . He shall live and become like a god . . . and he shall be hymned by the living ones every day continually and regularly forever."[24]

---

*It seems that other peoples, such as the West Africans and the Chinese, may have had traditions about this process; it is not the sort of knowledge given lightly to inquirers.

Conclusion

THE CIRCUITS OF THE ALL

Now unto the divine part in us the motions which are
kin are the thoughts and Circuits of the All. These must
every man follow that he may regulate the revolutions in
his head which were disturbed when the Soul was born in
the flesh; and by thoroughly learning the Harmonies and
Circuits of the All may make that which understandeth
like unto that which is understood.

PLATO, *TIMAEUS*

These are matters that deal with the most profound achievements of ancient wisdom. They have been disseminated and discussed at immense length over vast periods and areas. We are hardly liable, therefore, to be consistent and thorough in any attempt to expound them in a manageable form. Indeed, even Plato was aware of the unlikeliness of answers; consistency, he suggested, was a goal hardly within reach: "If in our treatment of a great host of matters regarding the gods and the generation of the universe, we prove unable to give accounts that are always in all respects self-consistent and perfectly exact, be not thou surprised. Rather we should be content if we are able to furnish accounts that are inferior to none in likelihood."[1]

Only by remaining "less than serious" can we, according to Plato,

220

reach the limits of science and retain the power of mystery. Neverthe-less, as Santillana reminds us, Plato had a warning to all who encountered his work: "For Plato not only has put into his piece [the *Timaeus*] all the science he can obtain, he has entrusted to it reserved knowledge of grave import, received from his archaic ancestors, and he soberly adjures the reader not to be too serious about it, nor even cultural in the modern sense, but to understand it if he can. The scholar is already in a hopeless tangle, and Lord help him."[2]

My intention in this book has been to uncover the way in which traditions of spiritual thought all find their original inspiration in the patterning that the human imagination discerns in the cycles of the stars. I hope the "probabilities," as Plato calls them, that I have adduced here are up to his required standard, and that I have put forward a convincing enough argument to suggest that the consistency is of such an order as to make them in fact far more likely than almost any others.

Among the "many opinions about the generation of the universe" that we have seen, one fundamental spiritual image emerges, that of the severance of heaven and earth. This image has been shown to have its origin in the recognition of the effects of the precession of the equinoxes, the understanding of the dynamic interaction of the "terrestrial" and "celestial" spheres. Gradual though it may have been, this recognition transformed the notion of the relationship that is said to have existed previously between the occupants of these two spheres, between Men and gods. More than anything else, it was this transformed relationship that supplied the architecture for the phenomenon that we experience as meaning.

For the traditions that Plato inherited, even the creation itself was to be regarded as a product of this generative division.* The familiar concepts of sin, of redemption, and of the soul and its immortality

---

*"It was already disputed in antiquity whether the notion that the universe had a first moment is or is not a part of the myth. . . . The dispute continues, the balance of opinion being that it is" (I. M. Crombie, *An Examination of Plato's Doctrines* 2, Sydney: Law Book Company of Australasia, 1963, 153).

have likewise been derived from this model. We have also seen how the social structures of traditional societies are representations of these notions. In short, it was the recognition of the shifts in cosmic order caused by the phenomenon of the precession that laid down the fundamental architecture of the human spirit.

In the Interlude chapter of this book, I suggested that the "Spirit" might be regarded as the "organ" of the brain that responds to the experience of resonance, the recognition of patterns and the patterns between patterns, what Bateson called "the pattern which relates." It seems to me that no better encompassing definition of love exists than this, the experience of this resonance. It was perhaps the emotional experience of these patterns of relatedness that first obliged our ancestors to explore the stars, a sense that the most profound pattern to which their spirit resonated must lie there. They were motivated, in fact, by the overpowering sense of love.

The discovery of the effects of the precession on cosmic order, of the corruptible nature of the pattern that they had felt embodied this love, must have changed the way that relatedness was experienced, and hence transformed the notion of love from one of immediate obviousness, of "being at one with nature"—a state of grace, if you like—into one of longing. Human relatedness became characterized by a shared sense of separation and a yearning for reconciliation in the eternal.

Today, the view through the kaleidoscope of human observation is once again causing the patterns of cosmology to shift, to expand into a much vaster perspective. Through the language of mathematics and astronomy, during our lifetime we have seen new patterns emerge whose significance is as great as, if not greater than, that felt by the ancients who first watched the cycle of the precession alter their notion of the universal forever. Anyone who has attempted to read a new sense of meaning into this shifting cosmos must still be reeling.

For the Zoroastrians, the duality of the universe that ensued from the discovery of the imperfections of creation, the rivalry of the divided heavens, was expressed as the confrontation between love and strife. Love was seen as the power that ordered and united, strife as the power that disordered and dissolved, and whose weapon was the "ignorance of the right order of things."

If, in the face of the contemporary cosmos, we introduce love as the mythic image of relatedness, then strife (the quality that, in other times, we might have called "evil") becomes truly the result of "the ignorance of the right order of things," the kind of response to the universe that arises when personal experience, for whatever reason, fails to confirm any sensation of universal relatedness. The spirit, faced with the collapse of the traditional cosmos, is all too prone to fall prey to the power of strife, to see in the universe only decay, disorder, and disruption, and to reverberate with that sense of dissolution. This is the ultimate expression of isolation, of desolation: a spirit without any sense of pattern.

To avoid such a fate, Plato's advice is that we strive to "make that which understandeth like unto that which is understood." For Plato, "that which is understood" (otherwise translated as "The Known") referred to the "Circuits of the All," the cycles of the cosmos, made manifest in the night sky, and "that which understandeth" ("the Knower") was the individual "circuits of the soul," bound in the "river of the body." The bringing of these two into harmony was the source of meaning for the human condition.

The dramatic changes in understanding the nature of the universe during the past century have produced an "All" whose circuits are vastly more complex than either mythological astronomy or Plato himself imagined. The precession of the equinoxes, which we have shown to have played the crucial role in determining our notions of creation, sin, and the immortal soul—ideas that have been with us since the beginning of time—is now seen as a minor result of local forces, of little consequence in the vastness of an ever-expanding cosmos. At the same time, our knowledge has expanded inward into the activities of the mind as they are revealed in the interactions of neural networks; more and more of the chemical and electrical processes that constitute "the circuits of the soul" can be observed and described.

The consequence of this escalation of understanding, however, has been the breakdown in our ability to achieve the harmonization of these two worlds, the external and the internal. It seems that as we learn more about the "understander" and the "understood," we know less about how to relate them. As a result, we are faced with some seriously challenging questions about our notions of spirituality and the sense of meaning they once generated.

It takes a bold poetic spirit indeed both to gaze out into the expanding cosmos and to probe inward into the paradox of consciousness in search of meaning; these are worlds more mysterious than any described by myth, worlds that stretch the imagination to its very limits. Nevertheless, it is by exploring the frontiers of the relationship between these two, cosmos and consciousness, that a new sense of meaning is slowly crystallized. The task is perhaps no greater than that encompassed by those ancient poets who, faced with the dissolution of an earlier myth, rebuilt the architecture of the spirit.

Appendix

THE RAISING OF THE SKY

The Oceanic myths of separation recorded by R. B. Dixon in *Oceanic Mythology* include some detailed descriptions of the separation of heaven and earth. They also give an enlightening view into the mode of expression of the mythical world that we have been exploring. They are reproduced here from the Internet site at www.sacred-texts.com/pac/om/om06.htm#fr_77.

According to the New Zealand belief, Rangi, the Sky Father, felt love for Papa-tu-a-nuku ("The Earth"), who lay beneath him, so he came down to Papa. At that time "absolute and complete darkness prevailed; there was no sun, no moon, no stars, no clouds, no light, no mist—no ripples stirred the surface of ocean; no breath of air, a complete and absolute stillness."[1] And Rangi set plants and trees to cover the nakedness of Papa, for her body was bare, placing insects of all kinds appropriate to the various sorts of vegetation, and giving their stations to the shellfish and the crabs and various sorts of living things. Then Rangi clave unto Papa, the Earth Mother, and held her close in his embrace, and as he lay thus prone upon Papa, all his offspring of gods which were born to him, both great and small,[2] were prisoned beneath his mighty form and lived cramped and herded together in darkness. "Because Rangi-nui over-laid and completely covered Papa-tua-nuku, the growth of all things could not mature, nor could anything bear fruit . . . they were in an unstable condition, floating about the Ao-pouri [the world of darkness], and this was their appearance: some were crawling, . . . some were upright with arms held up . . . some lying on their sides . . . some on their backs, some were stooping, some with their heads bent down, some with legs drawn up . . . some kneeling . . . some feeling about in the dark . . . they were all within the embrace of Rangi and Papa."[3] So for a long time the gods dwelt in darkness, but at last the desire came to them to better their condition, and for this purpose they planned to lift Rangi on high. The version of this myth of the raising of the sky, given by Sir George Grey,[4] is one of the classics of Polynesian mythology, and deserves to be quoted almost in full.

"Darkness then rested upon the heaven and upon the earth, and they still both clave together, for they had not yet been rent apart; and the children they had begotten were ever thinking amongst themselves what might be the difference between darkness and light; they knew that beings had multiplied and increased, and yet light had never broken upon them, but it ever continued dark. . . . At last the beings who had been begotten by Heaven and Earth, worn out by the continued darkness, consulted among themselves, saying, 'Let us now determine what we should do with Rangi and Papa, whether it would be better to slay them or to rend them apart.' Then spake Tu-matauenga, the fiercest of the children of Heaven and Earth, 'It is well, let us slay them.'

"Then spake Tane-mahuta, the father of forests and of all things that inhabit them, or that are constructed from trees, 'Nay, not so. It is better to rend them apart, and to let the heaven stand far above us, and the earth lie under our feet. Let the sky become a stranger to us, but the earth remain close to us as our nursing mother.' The brothers all consented to this proposal, with the exception of Tawhiri-ma-tea, the father of winds and storms, and he, fearing that his kingdom was about to be overthrown, grieved greatly at the thought of his parents being torn apart. Five of the brothers willingly consented to the separation of their parents, but one of them would not agree to it . . .

"But at length their plans having been agreed on, lo, Rongo-ma-tane, the god and father of the cultivated food of man, rises up, that he may rend apart the heavens and the earth; he struggles, but he rends them not apart. Lo, next Tangaroa, the god and father of fish and reptiles rises up, that he may rend apart the heavens and the earth, but he rends them not apart. Lo, next Haumia-tikitiki, the god and father of the food of man which springs up without cultivation, rises up and struggles, but ineffectually. Lo, then, Tu-matauenga, the god and father of fierce human beings, rises up and struggles, but he, too, fails in his efforts. Then, at last, slowly uprises Tane-mahuta, the god and father of forests, of birds, and of insects, and he struggles with his parents; in vain he strives to rend them apart with his hands and arms. Lo, he pauses; his head is now firmly planted on his mother the earth, his feet he raises up and rests against his father the skies, he strains his back and limbs with mighty effort. Now are rent apart Rangi and Papa, and with cries and groans of woe they shriek aloud, 'Wherefore slay you thus your parents? Why commit you so dreadful a crime as to slay us, as to rend your parents apart?' But Tane-mahuta pauses not, he regards not their shrieks and cries; far, far beneath him he presses down the earth; far, far above him, he thrusts up the sky . . .

"Up to this time, the vast Heaven has still ever remained separated from his spouse the Earth. Yet their mutual love still continues—the soft warm sighs of her loving bosom still ever rise up to him, ascending from the woody mountains and valleys, and men call these mists; and the vast Heaven, as he mourns through the long nights his separation from his beloved, drops frequent tears upon her bosom, and men seeing these, term them dewdrops."

Another Maori[5] version introduces several other elements:

"Raki, though speared by Takaroa, still adhered to the top of Papa; and Raki said to Tane and his younger brothers, 'Come and kill me, that men may live.' Tane said, 'O old man! how shall we kill you?' Raki said, 'O young man! lift me up above, that I may stand separate; that your mother may lie apart from me, that light may grow on you all.' Then Tane said to Raki, 'O old man! Rehua shall carry you.' Raki answered Tane and his younger brothers, 'O young men! do not let me be carried by your elder brothers only, lest my eyes become dim. Rather all of you carry me above, that I may be elevated," that light may dawn on you.' Tane said to him, 'Yes, O old man! your plan is right—that light may grow into day.' Raki said to Tane, 'It is right, O Tane! that I be taken and killed [separated from my wife], that I may become a teacher to you and your younger brothers, and show you how to kill. If I die, then will light and day be in the World.' Tane was pleased with the reasons why his father wished them to kill him; and hence Tane said to another branch of the offspring of Raki. . . . 'Tread on Papa, tread her down; and prop up Raki, lift him up above . . . that the eyes of Raki, who is standing here, may be satisfied.' . . . Now, this was the origin of the heaven. It was made by Tane and admired by him, and he uttered the words of his prayer to aid Rehua to carry their father above . . . Tane now took Raki on his back; but he could put Raki no higher. Raki said to Tane, 'You too, you and your younger brother [Paia] carry me.' Then Paia prayed his prayer, and said:

> 'Carry Raki on the back.
> Carry Papa.
> Strengthen, O big back of Paia,
> Sprained with the leap at Hua-rau.'

"Now, Raki was raised with the aid of this prayer, and spoke words of *poroporoaki* [farewell] to Papa, and said, 'O Papa! O! you remain here. This will be the [token] of my love to you; in the eighth month

I will weep for you.' Hence the origin of the dew, this being the tears of Raki weeping for Papa. Raki again said to Papa, 'O old woman! live where you are. In winter I will sigh for you.' This is the origin of ice. Then Papa spoke words of farewell to Raki, and said, 'O old man! go, O Raki! and in summer I also will lament for you.' Hence the origin of mist, or the love of Papa for Raki.

"When the two had ended their words of farewell, Paia uplifted Raki, and Tane placed his *toko* [pole] . . . between Papa and Raki. Paia did likewise with his *toko* . . . Then Raki floated upwards, and a shout of approval was uttered by those up above, who said, 'O Tu of the long face, lift up the mountain.' Such were the words shouted by the innumerable men [beings] from above in approval of the acts of Tane and Paia; but that burst of applause was mostly in recognition of Tane's having disconnected the heaven, and propped up its sides, and made them stable. He had stuffed up the cracks and chinks, so that Raki was completed and furnished, light arose and day began."[6]

Similar but briefer versions of this same myth are found in the Chatham Islands,[7] where the raising of the heavens was done by a being called "Heaven-Propper," the sky being lifted upon ten pillars, set one above the other. In the Cook Group,[8] the raiser of the heavens was Ru. Originally the heavens were low, so low that they rested on the broad leaves of certain plants, and in this narrow space all the people of this world were pent up, but Ru sent for the gods of night and the gods of day to assist him in his work of raising the sky. He prayed to them, "Come, all of you, and help me to lift up the heavens." And when they came in answer to his call, he chanted the following song:

> "O Son! O Son! Raise my son
> Raise my son!
> Lift the Universe! Lift the Heavens!
> The Heavens are lifted,
> It is moving!
> It moves,
> It moves!"

The heavens were raised accordingly, and Ru then chanted the following song to secure the heavens in their place:

> *"Come, O Ru-taki-nuhu,*
> *Who has propped up the Heavens.*
> *The Heavens were fast, but are lifted.*
> *The Heavens were fast, but are lifted.*
> *Our work is completed."*[9]

This conception, that the sky was originally low, resting on the leaves of plants, is also found in the Society Group,[10] where Ru is again the deity by whose aid the task of raising the heavens was accomplished. It likewise occurs in Samoa,[11] and in somewhat similar form in the Union Group,[12] whereas in Hawaii the incident of the separation of heaven and earth is referred to but vaguely and seems to play a very insignificant part in the beliefs of the people.[13]

It will be observed that the idea of a Sky Father and Earth Mother, so characteristic in New Zealand, is lacking in central Polynesia. What is said is merely that once the sky was very low, and that one of the deities raised it to its present position. Now this form of the myth appears in the New Hebrides,[14] where the heaven was said originally to have been so low that a woman struck it with her pestle as she was pounding food, whereupon she angrily told the sky to rise higher, and it did so. Almost identically the same type appears in the Philippines,[15] and the simple theme of raising the heavens, which once were low, is frequent in several other parts of Indonesia as well as in the intervening area of Micronesia. It would seem, therefore, that the Maori form of the myth represents a special or locally developed form of this widespread theme, which reaches back almost without a break from central Polynesia to Indonesia.

There may well be a great deal to be learned concerning movements of people and their historical contacts and relationships by studying these differences. The basic import of the story remains unchanged, however: once the sky lay close to the earth, until for some reason it came to be lifted up, propped apart, and firmly established. And that was the end of intimacy between heaven and earth, and the beginning of time.

APPENDIX NOTES

1. S. P. Smith, "The Lore of the Whare-wananga; or Teachings of the Maori College on Religion, Cosmogony and History," *Memoirs Polynesian Society* iii (1913): 117.

2. The number of these is given as seventy; see Smith, 1913, p. lig.

3. S. P. Smith, "The Lore of the Whare-wananga," 117.

4. W. Grey, "Some Notes on the Tannese," in *Austr. Assoc. Adv. Sci.* iv. (1892): 645–80, 117.

5. J. White, *The Ancient History of the Maori, His Mythology and Traditions*, 4 vols, (Wellington: G. Didsbury, 1886), i, 46ff.

6. For other Maori versions see White, i. 25, 26, 52, 138, 141, 161; also E. Best, "Notes on Maori Mythology," *Journal of the Polynesian Society* viii (1899): 93–121; J. F. Wohlers, "Mythology and Traditions of the Maori in New Zealand," *Proceedings New Zealand Institute* vii (1874): 3–53; E. Shortland, *Maori Religion and Mythology, with Translations of Traditions* (London, 1882), 20; Smith, 121, 314.

7. A. Shand, "The Moriori People of the Chatham Islands; Their Traditions and History," *Journal of the Polynesian Society* iii (1894): 76–92, 121–33, 187–98.

8. J. Pakoti, "The First Inhabitants of Aitutaki: The History of Ru," *Journal of the Polynesian Society* iv (1895): 65–70, 66.

9. For other versions, see W. W. Gill, *Myths and Songs from the South Pacific* (London: Henry S. King & Co., 1876), 59, 71; S. P. Smith, "Hawaiki; the Whence of the Maori," *Journal of the Polynesian Society* viii (1899): 1–49, 64. These, however, ascribe at least part of the task to Maui. See infra, pp. 50ff.

10. W. Ellis, *Polynesian Researches, during a Residence of nearly Eight Years in the Society and Sandwich Islands,* 4 vols. (New York: Kessinger Publishing, 1840), i, 100; J. A. Moerenhout, *Voyages aux îles du Grand Océan,* 2 vols. (Paris: Arthus Bertrand, 1837), i, 446.

11. A. Bastian, *Die samoanische Schöpfungssage und Anschliessendes aus der Südsee* (Berlin, 1894), 32; J. Fraser, "Some Folk-Songs and Myths from Samoa," *Journal of the Royal Society of New South Wales* xxv (1891): 70–86, 96–146, 242–86, 266; G. Turner, *Nineteen Years in Polynesia* (London: Jonh Snow, 1861), 245; cf. also S. P. Smith, "The Traditions of Nieue-fekai," *Journal of the Polynesian Society* xii (1903): 22–31, 85–119.

12. G. Turner, *Samoa a Hundred Years Ago and Long Before* (London: Macmillan and Co., 1884), 283.
13. D. Malo, *Hawaiian Antiquities* (Honolulu: Bishop Museum Press, 1903), 36.
14. D. Macdonald, "Efate, New Hebrides," *Austr. Assoc. Adv. Sci.* iv (1892): 720–35.
15. H. O. Beyer, "Origin Myths among the Mountain Peoples of the Philippines," *Philippine Journal of Science* sect. D, viii (1913): 85–118; L. W. Benedict, "Bagobo Myths," *Journal of American Folk-Lore* xxvi (1913): 13–63; Luzon (Ifugao), Beyer, 105.

NOTES

INTRODUCTION: MAKING SENSE

1. Giorgio de Santillana and Hertha von Dechend, *Hamlet's Mill,* 389.
2. At www.janeresture.com/banab/creation.htm (10/01/2006).
3. C. W. Peck, *Australia Legends,* 1925. At www.sacred-texts.com (10/01/2006).
4. Wendy O'Flaherty, *The Rig Veda,* 213, Hymn 7.86.
5. David Freidel, Linda Schele, and Joy Parker, *Maya Cosmos,* 115.
6. Mircea Eliade, *Patterns of Comparative Religion,* 184.
7. Santillana and von Dechend, *Hamlet's Mill.*
8. Ibid., viii.
9. Ibid., vii.
10. Ibid., viii.
11. William Sullivan, *The Secret of the Incas,* 9–10.
12. E. C. Krupp, *Echoes of the Ancient Skies,* 1.
13. Daniel C. Dennett, *Consciousness Explained,* 178.

The Myth of Creation

OUT OF THE ABYSS

1. Stephen Hawking, *A Brief History of Time,* 46.
2. O'Flaherty, *Rig Veda,* 25, Hymn 10.129, i.
3. B. van de Walle, "Egypt: Syncretism and State Religion," *Larousse World Mythology* (London: Hamlyn, 1973), 30.
4. Genesis 1:2.
5. Mircea Eliade, *Essential Sacred Writings from Around the World,* 86.
6. George Buhler, trans., *The Laws of Manu,* 1, 5. At www.sacred-texts.com/hin/manu.htm (10/12/2006).
7. Hawking, *Brief History of Time,* 10–11.

8. Apollonius Rhodius, *Argonautica,* trans. R. C. Seaton, 1912. At www.sacred-texts.com/cla/argo/index.htm (10/17/2006). Quoted in M. Detienne and J.-P. Vernant, *Cunning Intelligence in Greek Culture and Society,* trans. J. Lloyd (Sussex, England: Harvester Press, 1978), 49–50.

9. Quoted in Eliade, *Essential Sacred Writings,* 98.

10. Eliade, *Patterns of Comparative Religion,* 212.

11. C. J. Jung, *Psychology and Alchemy,* trans. R. F. C. Hull (London: Routledge and Kegan Paul, 1968), 25.

12. Hesiod, *Theogony,* 728 and 738. At www.sacred-texts.com. Quoted in Detienne and Vernant, *Cunning Intelligence in Greek Culture and Society,* 167.

13. George Buhler, trans., *Laws of Manu.*

14. Vishnu Purana, 34. Quoted in Eliade, *Essential Sacred Writings,* 42.

15. Anthony Christie, *Chinese Mythology,* 49.

16. Eliade, *Patterns of Comparative Religion,* 413.

17. Coffin Texts 714. Quoted in Eliade, *Essential Sacred Writings,* 24.

18. Mircea Eliade, *The Myth of the Eternal Return,* 212.

19. Quoted in W. K. C. Guthrie, *A History of Greek Philosophy,* 301.

A COMPASS UPON THE DEEP

1. Apollonius Rhodius, *Argonautica.*

2. Detienne and Vernant, *Cunning Intelligence in Greek Culture and Society,* 146.

3. Santillana and von Dechend, *Hamlet's Mill,* 4.

4. E. A. Wallis Budge, *The Book of the Dead* (London: Arkana, 1989), 13, n. 1.

5. Ibid., 67, ch. 15, Hymn and Litany to Osiris.

6. Jung, *Psychology and Alchemy,* 318.

7. O'Flaherty, *Rig Veda,* 25, Hymn 10.129, 1, 4–5.

8. Job 38:5.

9. Mircea Eliade, *A History of Religious Ideas,* 207.

10. *Popol Vuh,* ch. 1. Quoted in Eliade, *Essential Sacred Writings,* 93.

11. Mircea Eliade, *Australian Religion* (Ithaca, N.Y.: Cornell University Press, 1973), 67 ff.

12. *Mahabharata,* 12.290. Quoted in Margaret Stutely and James Stutely, *A Dictionary of Hinduism,* 206, n. 2, Narayana.

13. Ibid., quoting Maha Narayana Upanishad.

14. R. T. Rundle Clark, *Myth and Symbol in Ancient Egypt,* 139–40.

15. Jung, *Psychology and Alchemy,* 382.

16. R. C. Zaehner, *The Dawn and Twilight of Zoroastrianism,* 201.

17. Stutely and Stutely, *Dictionary of Hinduism,* 57.

18. O'Flaherty, *Rig Veda,* 77, Hymn 1.164, 2, 11.

19. Jung, *Psychology and Alchemy,* 381.

20. Stutely and Stutely, *Dictionary of Hinduism,* 57.
21. O'Flaherty, *Rig Veda,* 268, Hymn 10.85, 16.

HOW THE EARTH WAS MADE

1. Eliade, *Patterns of Comparative Religion,* 212.
2. Ibid.
3. Ibid.
4. R. Tsunoda et al., *Sources of Japanese Tradition,* vol. 1, 25.
5. M. Soymie, "China: The Struggle for Power," in *Larousse World Mythology,* 274.
6. Santillana and von Dechend, *Hamlet's Mill,* 413.
7. Freidel, Schele, and Parker, *Maya Cosmos,* 53.
8. John G. Neihardt, *Black Elk Speaks,* 22, 27.
9. A. M. Panoff, "Oceania: Society and Tradition," in *Larousse World Mythology,* 507.
10. Psalms 48:2.
11. Freidel, Schele, and Parker, *Maya Cosmos,* 140.
12. R. T. Rundle Clark, *Myth and Symbol in Ancient Egypt.*
13. K. Langloh Parker, *The Euahlay Tribe: A Study of Aboriginal Life in Australia.* At www.sacred-texts.com/aus/tct/index.htm (10/04/2006).
14. James Mooney, "Myths of the Cherokee," *Nineteenth Annual Report of the Bureau of American Ethnology 1897–98,* part 1, 1900. At www.sacred-texts.com/nam/cher/motc/index.htm (10/04/2006).
15. Plato, *Republic,* "The Vision of Er."
16. Freidel, Schele, and Parker, *Maya Cosmos,* 130.
17. Richard Laurence, trans., *The Book of Enoch the Prophet,* ch. 18, 3.
18. O'Flaherty, *Rig Veda,* 203, Hymn 1.160.

THE MEASURES OF CREATION

1. *Popol Vuh,* trans. D. Tedlock. Quoted in Freidel, Schele, and Parker, *Maya Cosmos,* 107.
2. Laurence, *Book of Enoch,* ch. 60, 2–6.
3. Job 38:4–5.
4. Michael Loewe, *Ways to Paradise,* 74.
5. Freidel, Schele, and Parker, *Maya Cosmos,* 57.
6. Eliade, *Essential Sacred Writings,* 159.
7. Eliade, *Myth of the Eternal Return,* 52.
8. Freidel, Schele, and Parker, *Maya Cosmos,* 122.
9. R. Bastide, "Africa: Magic and Symbolism," in *Larousse World Mythology,* 537.
10. Krupp, *Ancient Skies,* 241.

The Myth of Corruption

THE GOLDEN RULE

1. Santillana and von Dechend, *Hamlet's Mill*, 83.
2. Ibid., 135.
3. Ibid., 134.
4. Ibid.
5. Job 3:84.
6. Santillana and von Dechend, *Hamlet's Mill*, 135.
7. Ibid., 136, n. 49.
8. Eliade, *Patterns of Comparative Religion*, 62.
9. Santillana and von Dechend, *Hamlet's Mill*, 135.
10. Marie-Louise von Franz, *Number and Time*, 159.
11. Zaehner, *Zoroastrianism*, 134.
12. Eliade, *History of Religious Ideas*, 32.
13. Genesis 3:8.
14. J. G. Frazer, *The Belief in Immortality* (London: Macmillan, 1913), 63.
15. Eliade, *Essential Sacred Writings*, 156.

PRIDE AND FALL

1. Eliade, *Patterns of Comparative Religion*, 211.
2. Livio C. Stechinni, "Notes on the Relation of Ancient Measures to the Great Pyramid," in Peter Tompkins, *Secrets of the Great Pyramid*, 337.
3. von Franz, *Number and Time*, 159.
4. Mahabharata. Quoted in Santillana and von Dechend, *Hamlet's Mill*, 152.
5. Wallis Budge, *Book of the Dead*, 596, ch. 175.
6. Eliade, *History of Religious Ideas*, 241.
7. Enuma Elish. Quoted in Eliade, *Essential Sacred Writings*, 98.
8. Ibid., 99.
9. Isaiah 14:12–13.
10. Psalms 82:6–8.
11. Ibid., 58:1.
12. Joseph Campbell, *The Hero with a Thousand Faces*, 22.
13. Wallis Budge, *Book of the Dead*, ch. 177.
14. Enuma Elish. Quoted in Eliade, *Essential Sacred Writings*, 106.
15. Zaehner, *Zoroastrianism*, 134, 315.
16. Ibid.
17. Ibid.
18. David Bohm, *Wholeness and the Implicate Order* (London: Routledge, 1983), 168.
19. Bhagavadgita. Quoted in von Franz, *Number and Time*, 255.

THE SEPARATION OF HEAVEN AND EARTH

1. Robert Graves, *The White Goddess* (London: Faber and Faber, 1961), 49.
2. Isaiah 22:25, 24:18,19, 23.
3. Rundle Clark, *Myth and Symbol in Ancient Egypt,* 139–40.
4. Voluspa, 59–60. Quoted in Eliade, *Essential Sacred Writings,* 126.
5. Revelation 20:1.
6. R. B. Dixon, *Oceanic Mythology,* 1916, 15. At www.sacred-texts.com/pac/om/ om06.htm#fr,%2030 (10/05/2006).
7. Anaximander. Quoted in Guthrie, *Greek Philosophy,* vol. 1, 100.
8. Eliade, *History of Religious Ideas,* 63.
9. Charles, *The Book of Enoch,* chapter 18.
10. Skanda Purana. Quoted in O'Flaherty, *Hindu Myths,* 240–43.
11. Markandeya Purana. Quoted in O'Flaherty, *Hindu Myths,* 248.
12. Soymie, "China: The Struggle for Power," in *Larousse World Mythology,* 277.
13. Laurence, *Book of Enoch,* book 5, ch. 64.
14. Krupp, *Ancient Skies,* 147.
15. Enuma Elish. Quoted in Eliade, *Essential Sacred Writings,* 99.
16. Ibid.
17. Stephen Skinner, *Terrestrial Astrology,* 55.
18. Quoted in Santillana and von Dechend, *Hamlet's Mill,* 134.
19. Milton, *Paradise Lost,* book 10, lines 668–79.

NIGHTS IN THE GARDEN OF EDEN

1. Stutely and Stutely, *Dictionary of Hinduism,* "cakra," 57, RV 10.85, 16.
2. Orphic fragment. Quoted in Santillana and von Dechend, *Hamlet's Mill,* 134.
3. Robert Bauval and Graham Hancock, *Keeper of Genesis.*
4. Santillana and von Dechend, *Hamlet's Mill,* 258.
5. Freidel, Schele, and Parker, *Maya Cosmos,* 76–77.
6. Ibid., 71.
7. Santillana and von Dechend, *Hamlet's Mill,* 227.
8. Ibid., 259.
9. R. O. Faulkner, "Invocation of the Ladder to the Sky," *The Ancient Egyptian Pyramid Texts,* 165, Utterance 478.
10. Ibid., ch. 149.
11. Freidel, Schele, and Parker, *Maya Cosmos,* 105–6.
12. Frazer, *The Belief in Immortality,* 72.
13. Freidel, Schele, and Parker, *Maya Cosmos,* 105–6.
14. Ibid., 255.
15. T. G. H. Strehlow, *Aranda Traditions* (New York, London: Johnson Reprint, 1968).
16. Krupp, *Echoes of the Ancient Skies,* 2.

ONE AND ALL

1. Plato, *Timaeus.*
2. Arthur Fairbanks, ed. and trans., "Parmenides Fragments and Commentary," in *The First Philosophers of Greece* (London: Keegan Paul, Trench, Trubner, 1898), 86–135. At history.hanover.edu/texts/presoc/parmends.html (10/10/2006).
3. Vishnu Purana, in O'Flaherty, *Hindu Myths,* 186.
4. Eliade, *Patterns in Comparative Religion,* 188.
5. Irenaeus, *Against Heresies,* book 1, ch. 17, in D. D. Alexander Roberts and James Donaldson, *Anti-Nicene Fathers,* vol. 1. At www.sacred-texts.com/chr/ecf/001/0010714.htm (10/10/2006).
6. Romans 8:20–23.
7. von Franz, *Number and Time,* 62, n. 7.
8. Detienne and Vernant, *Cunning Intelligence in Greek Culture and Society,* 88.
9. Plato, Timaeus, in *Plato,* vol. 9, translated by R. G. Bury (Cambridge, Mass.: Harvard University Press, 1929).
10. von Franz, *Number and Time,* 62.

BEING AND BECOMING

1. Eliade, *Essential Sacred Writings,* 155ff.
2. Plato, *Timaeus.*
3. Ibid.
4. Ibid.
5. Ibid.
6. Plato, *Politikos* ("Statesman").
7. Bohm, *Wholeness and the Implicate Order,* 19–24.
8. Ibid.
9. Loewe, *Ways to Paradise,* 77.
10. Santillana and von Dechend, *Hamlet's Mill,* 345.
11. Eliade, *Patterns of Comparative Religion,* 184.

Interlude

SPIRITUAL ARCHITECTURE

1. Santillana and von Dechend, *Hamlet's Mill,* 7.
2. Ibid., ix.
3. Ibid., ix.
4. Ibid., 346.
5. Ibid., 332ff.
6. Bastide, "Africa: Magic and Symbolism," 536–37.
7. C. Lévi-Strauss, *La pensée sauvage,* 1962 (translated 1966).
8. Bastide, "Africa: Magic and Symbolism," 536–42.
9. Krupp, *Ancient Skies,* 319.

10. Ibid.

11. Ibid., 318.

12. O'Flaherty, *Rig Veda*, 16–17.

The Myth of Reunion

THE MOTHER OF ALL

1. Plato, *Timaeus*, 125.

2. Ibid., 123.

3. Ibid., 119.

4. Elaine Pagels, *The Gnostic Gospels*, 50.

5. Ibid., 51.

6. von Franz, *Number and Time*, 171.

7. Quoted in R. H. Charles, *Apocrypha and Pseudepigraphia*.

8. Detienne and Vernant, *Cunning Intelligence in Greek Culture and Society*, 138.

9. Plato, *Timaeus*, 119.

10. Eliade, *Patterns in Comparative Religion*, 18.

11. Merlin Stone, *Paradise Papers*, 9–10.

12. Ibid., 227.

13. Book of Wisdom, 8:8, in *The Jerusalem Bible* (London: Darton, Longman and Todd, 1966).

14. Detienne and Vernant, *Cunning Intelligence in Greek Culture and Society*, 137.

15. Wallis Budge, *Book of the Dead*, 4, n. 5.

16. Stechinni, "Ancient Measures," in Tompkins, *Secrets of the Great Pyramid*, 337.

17. Stutely and Stutely, *Dictionary of Hinduism*, "rita."

18. Eliade, *History of Religious Ideas*, 201.

19. Stone, *Paradise Papers*, 216.

20. Wallis Budge, *Book of the Dead*, ch. 17.92, ch. 17.145, 111.

21. Ibid., ch. 136.12, 409.

22. Stutely and Stutely, *Dictionary of Hinduism*, "svasti."

23. Plato, *Phaedrus*, trans. by Benjamin Jowett. At www.sacred-texts.com (10/16/2006).

24. Aetus. Quoted in Guthrie, *Greek Philosophy*, 234.

25. O'Flaherty, *Hindu Myths*, 274.

26. Ibid., 275.

27. Santillana and von Dechend, *Hamlet's Mill*, 381.

28. Plato, *Politikos*, trans. by Benjamin Jowett. At www.sacred-texts.com (10/16/2006).

29. Ibid.

NATURE AND CULTURE

1. C. Lévi-Strauss, *Introduction to a Science of Mythology*, vol. 4, 622.

2. Ibid., 499.

3. Jung, *Mysterium Coniunctionis* (London: Routledge, 1963), 159, fn. 336.

4. Genesis 25:24.

5. Lévi-Strauss, *Science of Mythology,* vol. 4, 622.

6. Ibid., 501.

7. Santillana and von Dechend, *Hamlet's Mill,* 322.

8. J. G. Frazer, *Myths of the Origin of Fire.*

9. Rig Veda, Hymn 5.13.6. Quoted in Santillana and von Dechend, *Hamlet's Mill,* 140.

10. Dixon, *Oceanic Mythology.*

11. Hesiod, *Theogony.*

12. A. Rees and B. Rees, *Celtic Heritage,* 156.

13. Ibid., 157–58, quoting A. K. Coomeraswamy, *The Rig Veda as Land-Nams-Bok* (London, 1935).

14. Campbell, *Hero with a Thousand Faces,* 98–99, 139–40.

15. Ibid., 140.

THE SAVIOR OF THE WORLD

1. Santillana and von Dechend, *Hamlet's Mill,* 265.

2. Ferdowsi, *Shahnameh,* translated by Reuben Levy (London: Arkana, 1990), 9.

3. Vishnu Purana. Quoted in Giorgio de Santillana and Hertha von Dechend, *Hamlet's Mill,* 407–8.

4. Rees and Rees, *Celtic Heritage,* 73.

5. Ferdowsi, *Shahnameh,* 20.

6. Stutely and Stutely, *Dictionary of Hinduism,* 88, "Gada."

7. Detienne and Vernant, *Cunning Intelligence in Greek Culture and Society,* 76.

8. Stutely and Stutely, *Dictionary of Hinduism,* 88, "vajra."

9. Manilius, *Astronomica,* translated by Thomas Creech. At www.cyberwitch.com/wychwood/observatory/manilius.htm (10/12/2006).

10. Richard Hinckley Allen, *Star Names: Their Lore and Meaning.*

HEAVEN ON EARTH

1. Margaret A. Murray, *The Divine King of England* (London: Faber and Faber, 1954).

2. Ferdowsi, *Shahnameh,* 11.

3. Ibid.

4. Rees and Rees, *Celtic Heritage,* 129.

5. Ibid., 74.

6. Eliade, *Myth of the Eternal Return,* 80.

7. Ibid., 29.

8. W. Marsham Adams, *The Book of the Master of Hidden Places* (Wellingborough, U.K.: Aquarian Press, 1980), 7–71.

9. Rees and Rees, *Celtic Heritage*, 146.
10. Eliade, *Patterns in Comparative Religion*, 462.
11. Ibid., 66–67.
12. O'Flaherty, *Rig Veda*, 267, Hymn 10.85.
13. Eliade, *Essential Sacred Writings*, 165.
14. Marcel Granet, *La pensée Chinoise*. Quoted in Eliade, *History of Religious Ideas*.
15. Eliade, *Essential Sacred Writings*, 170–2. Eliade's text is from Hans Scharer's *The Conception of God among a South Borneo People*, translated by Rodney Needham (The Hague, 1963), 94–97.
16. Eliade, *Myth of the Eternal Return*, 56.
17. Rees and Rees, *Celtic Heritage*, 169ff.
18. Eliade, *Myth of the Eternal Return*, 53.
19. D. Zahan, *Religion, Spirituality and Thought of Traditional Africa*, 66, 78, 80.
20. Eliade, *Myth of the Eternal Return*, 63.

BUILDING THE KINGDOM: AS ABOVE, SO BELOW

1. Eliade, *Myth of the Eternal Return*, 28.
2. R. Bastide, "Africa: Magic and Symbolism," in *Larousse World Mythology*, 540–42.
3. A. Sauvageot, "Finland-Ugria: Magic Animals," in *Larousse World Mythology*, 425.
4. Krupp, *Ancient Skies*, 236–37.
5. Zahan, *Traditional Africa*, 69.
6. Ibid., 72.
7. Ibid., 77.
8. R. W. Sloley, "Primitive Methods of Measuring Time, with Special Reference to Egypt," *Journal of Egyptian Archaeology* 57 (1931), 110–19. Quoted at www.egyptstudy/ostrcon/archives/GreenwallSummer2005.pdf (05/26/2007).
9. Krupp, *Ancient Skies*, 231.
10. Santillana and von Dechend, *Hamlet's Mill*, 220.
11. Krupp, *Ancient Skies*, 239.
12. Ibid., 241.
13. Eliade, *Essential Sacred Writings*, 161.
14. Krupp, *Ancient Skies*, 260.
15. Ibid.
16. Ibid., 270.
17. Rees and Rees, *Celtic Heritage*, 156.
18. Ibid.
19. Ibid., 159.
20. Ibid., 162.
21. Ibid., 163.
22. Neihardt, *Black Elk Speaks*, 22.

THE GREAT TRIBE

1. Rees and Rees, *Celtic Heritage*, 101ff.
2. Annie Lebeuf, "Le système classificatoire des Fali," in Zahan, *Traditional Africa*, 328ff.
3. Emile Durkheim, *The Elementary Forms of the Religious Life*, 146.
4. Ibid., 293.
5. Jackson, *The Symbol Stones of Scotland*, 86–87.
6. Durkheim, *Religious Life*, 141.
7. Ibid., 189.
8. Ibid., 291.
9. Ibid., 240.
10. W. Lloyd Warner, *A Black Civilization*. Quoted in Eliade, *Essential Sacred Writings*, 187.
11. Durkheim, *Elementary Forms of the Religious Life*, 265.
12. Strehlow, *Arandan Traditions*.
13. Sullivan, *The Secret of the Incas*, 241.
14. von Franz, *Number and Time*, 267.
15. Ibid., 268.
16. Lang, *Custom and Myth*, 150.
17. Ibid.
18. H. B. Alexander, *North American Mythology*. Quoted in Santillana and von Dechend, *Hamlet's Mill*, 309.
19. W. Lloyd Warner, *A Black Civilization*. Quoted in Eliade, *Essential Sacred Writings*, 187.
20. Durkheim, *Elementary Forms of the Religious Life*, 290.
21. Eliade, *Australian Religion*, 29.
22. Sullivan, *Secret of the Incas*, 57–58.
23. Plato, *Timaeus*, 41–42.
24. Ibid., 35.
25. Eliade, *Australian Religion*, 31.
26. Guthrie, *Greek Philosophy*, vol. 1, 201.
27. Hegemonius. Quoted in Jung, *Psychology and Alchemy*, 380, n. 110.
28. G. Dieterlen, "L'initiation chez les pasteurs Peul," in M. Fortes and G. Dieterlen, eds., *African Systems of Thought*, 314.
29. Plato, *Phaedo*, 69, translated by Benjamin Jowett at classics.mit.edu/plato/phaedo.html.
30. Zaehner, *Zoroastrianism*, 239.

THE WAY HOME

1. Rig Veda, Hymn 10.14. Quoted in Santillana and von Dechend, *Hamlet's Mill*, 304.

2. Stutely and Stutely, *Dictionary of Hinduism,* "pitris."

3. James Darmesteter, *Zend Avesta,* 15–18. Quoted in Eliade, *Essential Sacred Writings,* 389.

4. Matsaya Purana. Quoted in O'Flaherty, *Hindu Myths,* 181.

5. Ibid., 181.

6. Andrew George, trans., *The Epic of Gilgamesh,* 1.

7. W. W. Gill, *Myths and Songs from the South Pacific* (London: H. S. King, 1876). Quoted in Santillana and von Dechend, *Hamlet's Mill,* 316.

8. Rig Veda, Hymn 1.13, 6. Translated by R. T. H. Griffith; posted at www.sacred-texts.com/hin/rigveda/rv01013.htm.

9. Cicero, *Dream of Scipio.*

10. Macrobius. Quoted in Santillana and von Dechend, *Hamlet's Mill,* 242.

11. Santillana and von Dechend, *Hamlet's Mill,* 240.

12. Prashna Upanishad, 9, in R. C. Zaehner, trans., *Hindu Scriptures.*

13. Santillana and von Dechend, *Hamlet's Mill,* 243.

14. Pindar, *Olympian Odes,* 2.68, at www.perseus.tufts.edu.

15. Doresse, *The Secret Books of the Egyptian Gnostics,* 73.

16. Graham Hancock, *Heaven's Mirror,* 73.

17. Ibid., 76.

18. Wallis Budge, *Book of the Dead,* ch. 135, 378, n. 1.

19. Doresse, *Egyptian Gnostics,* 73.

20. Ibid., 79.

21. Jung, *Psychology and Alchemy,* 310–11.

22. The Ordinary of the Mass according to the Old Catholic Missal, oblation of the wine.

23. Doresse, *Egyptian Gnostics,* 76.

24. Budge, *Book of the Dead,* 411, rubric to ch. 136a.

CONCLUSION: THE CIRCUITS OF THE ALL

1. Santillana and von Dechend, *Hamlet's Mill,* 328.

2. Lewis-Williams, *The Mind in the Cave,* 131.

BIBLIOGRAPHY

Allen, Richard Hinckley. *Star Names: Their Lore and Meaning.* New York: Dover, 1963.

Armitage, Frederick. *The Old Guilds of England.* London: Weare, 1918.

Armstrong, John. *The Paradise Myth.* Oxford: Oxford University Press, 1969.

Bauval, Robert, and Graham Hancock. *Keeper of Genesis.* London: Heinemann, 1996.

Bennett, Florence M. *Religious Cults Associated with the Amazons.* New York: Columbia University Press, 1912. At www.sacred-texts.com/wmn/rca/index.htm.

Bohm, David. *Wholeness and the Implicate Order.* London: Routledge, 1980; paperback edition, 1983.

Brandon, S. G. F. *History, Time and Deity.* Manchester, U.K.: Manchester University Press, 1965.

Bromwich, Rachel. *Trioedd Ynys Prydein* 1961. Cardiff: University of Wales Press, 2006.

Brophy, T. G. *The Origin Map.* New York: iUniverse, 2002.

Brown, Robert. *Research into the Origins of the Primitive Constellations of the Greeks, Phoenicians and Babylonians.* 2 vols. London: Williams and Norgate, 1899/1900.

Bryant, Jacob. *A New System, or An Analysis of Ancient Mythology.* 3 vols. London: printed for J. Walker, W. J. and J. Richardson, 1807.

Campbell, Joseph. *The Hero with a Thousand Faces.* 1949. London: Sphere, 1975.

———. *The Masks of God.* 4 vols. 1962. New York: Penguin, 1976.

———, ed. *Mysteries: Papers from the Eranos Year Books.* Princeton, N.J.: Princeton University Press, 1979.

Charles, R. H. *Apocrypha and Pseudepigraphia.* Oxford: Oxford University Press, 1913.

———. *The Book of Enoch.* London: Society for Promoting Christian Knowledge, 1917.

Christie, Anthony. *Chinese Mythology*. London: Hamlyn, 1968.

Cooke, A. B. *Zeus: A Study in Ancient Religion*. Cambridge, U.K.: Cambridge University Press, 1914.

Crocker, Jon Christopher. *Vital Souls: Bororo Cosmology, Natural Symbolism and Shamanism*. Tucson: University of Arizona Press, 1985.

Crombie, I. M. *An Examination of Plato's Doctrines*. 2 vols. London: Routledge and Kegan Paul, 1967.

Darmesteter, J. *Zend Avesta*. Oxford: Oxford University Press, 1880.

Dennett, Daniel C. *Consciousness Explained*. London: Penguin Books, 1993.

Dixon, Roland B. *Oceanic Mythology*. London: Marshall Jones Company, 1916; www.sacred-texts.com.

Doresse, Jean. *The Secret Books of the Egyptian Gnostics*. New York: Viking, 1960.

Dowson, J. *A Classical Dictionary of Hindu Mythology*. London: Trubner, 1913.

Durkheim, Emile. *The Elementary Forms of the Religious Life*. Trans. by Joseph Ward Swain. London: Allen and Unwin, 1976.

Eliade, Mircea. *Essential Sacred Writings from Around the World*. New York: Harper and Row, 1977. Originally published as *From Primitives to Zen*, 1967.

———. *A History of Religious Ideas*. Chicago: University of Chicago Press, 1988.

———. *The Myth of the Eternal Return*. London: Routledge and Kegan Paul, 1955.

———. *Patterns of Comparative Religion*. Lincoln: University of Nebraska Press, 1996.

———. *Shamanism*. Princeton, N.J.: Princeton University Press, 1972.

Encyclopaedia Judaica. Jerusalem: Encyclopaedia Judaica, 1971.

Faulkner, Raymond O., trans. *The Ancient Egyptian Book of the Dead*. London: British Museum Publications, 1985.

———. *The Ancient Egyptian Pyramid Texts*. Warminster, U.K.: Aris and Phillips, 1969.

Fergusson, James. *Tree and Serpent Worship* or *Illustrations of Mythology and Art in India*. Whitefish, Mont.: Kessinger Publishing, 2004.

Fortes, M., and G. Dieterlen, eds. *African Systems of Thought*. London: Oxford University Press, 1969.

Foundation for the Advancement of Mesoamerican Studies, Inc., www.famsi.org.

Frazer, James G. *The Belief in Immortality and the Worship of the Dead*. 3 vols., 1913–22. London: Routledge Curzon, 1994.

———. *Myths of the Origin of Fire*. London: Macmillan, 1930.

Freidel, David, Linda Schele, and Joy Parker. *Maya Cosmos*. New York: Morrow, 1995.

Freund, Philip. *Myths of Creation*. London: Peter Owen, 1964.

Frobenius, Leo, and Douglas Fox. *African Genesis*. London: Faber and Faber, 1938.

Ganz, J., trans. *The Mabinogion*. London: Penguin, 1977.

George, Andrew, trans. *The Epic of Gilgamesh*. London: Allen Lane; Penguin, 1999.

Gimbutas, Marija. *The Gods and Goddesses of Old Europe*. London: Thames and Hudson, 1974.

Grant, Kenneth. *Cults of the Shadow*. London: Muller, 1975.

Grimal, P., ed. *Larousse World Mythology.* London: Hamlyn, 1973.

Guillaumont, A. *The Gospel According to Thomas.* London: Harper Collins, 1984.

Guthrie, W. K. C. *A History of Greek Philosophy.* 5 vols. Cambridge, U.K.: Cambridge University Press, 1962–65.

Hackin, J., et al. *Asiatic Mythology.* London: Harrap, 1932.

Hadland Davis, F. *Myths and Legends of Japan.* London: Harrap, 1912.

Hancock, Graham. *The Fingerprints of the Gods.* London: Heinemann, 1995.

———. *Heaven's Mirror.* London: Michael Joseph, 1998.

Hawking, Stephen. *A Brief History of Time.* London: Bantam, 1988.

Hugh-Jones, Stephen. *The Palm and the Pleiades: Initiation and Cosmology in Northwest Amazonia.* Cambridge, U.K.: Cambridge University Press, 1979.

Jackson, Anthony. *The Symbol Stones of Scotland.* Kirkwall, Orkney, Scotland: Orkney Press, 1984.

James, E. O. *Creation and Cosmology.* Leiden, the Netherlands: E. J. Brill, 1969.

———. *The Worship of the Sky God.* London: University of London Press, 1963.

Jaynes, Julian. *The Origin of Consciousness in the Breakdown of the Bicameral Mind.* London: Penguin, 1993.

Jung, C. J. *Mysterium Coniunctionis. Collected Works,* vol. 14. London: Routledge and Kegan Paul, 1980.

———. *Paracelsus as a Spiritual Phenomenon. Collected Works,* vol. 13. London: Routledge and Kegan Paul, 1980.

———. *Psychology and Alchemy. Collected Works,* vol. 12. London: Routledge and Kegan Paul, 1980.

Kirk, G. S. *Myth: Its Meaning and Function in Ancient and Other Cultures.* Cambridge, U.K.: Cambridge University Press, 1970.

Krupp, E. C. *Echoes of the Ancient Skies.* Oxford: Oxford University Press, 1994.

Lang, Andrew. *Custom and Myth.* London: Longmans, Green, 1885.

Laurence, Richard, trans. *The Book of Enoch the Prophet.* Oxford: Oxford University Press, 1838.

Lévi-Strauss, Claude. *Introduction to a Science of Mythology.* London: Cape, 1970.

Levy, Reuben, trans. *The Epic of the Kings: Shahnameh, the National Epic of Persia.* 1967. London: Arkana, 1990.

Lewis-Williams, David. *Cosmos in Stone.* Walnut Creek, Calif.: Newman & Littlefield, 2002.

———. *The Mind in the Cave.* London: Thames and Hudson, 2004.

Lindsay, Jack. *Origins of Astrology.* London: Muller, 1971.

Loewe, Michael. *Ways to Paradise.* London: Allen and Unwin, 1979.

Lönnrot, Elias. *The Kalevala.* Trans. by Keith Bosley. Oxford: Oxford University Press, 1989.

Macchioro, Vittorio D. *From Orpheus to Paul.* London: H. Holt, 1930.

MacKenzie, Donald A. *Myths of Crete and Pre-Hellenic Europe.* London: Milford, 1973.

Marsham Adams, W. *The Book of the Master of the Hidden Places.* Wellingborough, U.K.: Aquarian Press, 1980.

Massey, G. *A Book of the Beginnings.* London: Williams, 1881.

Mbiti, John S. *African Religions and Philosophy.* London: Heinemann, 1969.

Mead, G. R. S. *Fragments of a Faith Forgotten.* London: Theosophical Publications Society, 1900. At www.sacred-texts.com/gno/fff/index.htm (10/01/2006).

Mellaart, James. *Catal Huyuk.* London: McGraw-Hill, 1967.

Menner, R. J. *The Poetical Dialogues of Solomon and Saturn.* New York: Kraus Reprint, 1941.

Milton, John. *Paradise Lost.* 1667. London: Penguin, 1996.

Muller, Max. *Sacred Books of the East.* At www.sacred-texts.com (10/30/2006).

Murray, M. A. *The Divine King in England.* Brooklyn, N.Y.: AMS Press, 1980.

Neihardt, John G. *Black Elk Speaks.* Lincoln: University of Nebraska Press, 1979.

Noble, N., and A. Coomaraswamy. *Myths of the Hindus and Buddhists.* New York: Dover, 1967.

O'Flaherty, Wendy D. *Hindu Myths.* London: Penguin, 1975.

———. *The Rig Veda.* London: Penguin, 1981.

Omasade Awolalu, J. *Yoruba Beliefs and Sacrificial Rites.* New York: Longmans, 1979.

Pagels, Elaine. *The Gnostic Gospels.* London: Vintage, 1981.

Paget, R. F. *In the Footsteps of Orpheus.* New York: Roy, 1968.

Parke, H. W., and D. Wormell. *A History of the Delphic Oracle.* Oxford: Blackwell, 1939.

Pausanius. *Guide to Greece.* 2 vols. Trans. by P. Levi. London: Penguin, 1979.

Phene, J. S. *On Prehistoric Traditions and Customs Connected with Sun and Serpent Worship.* London: R. Hardwicke, 1875.

Plato. *Timaeus,* in *Plato,* vol. 9. Trans. by R. G. Bury. Cambridge, Mass.: Harvard University Press, 1929.

Platt and Brett, eds. *The Lost Books of the Bible and the Forgotten Books of Eden.* Cleveland: World Publishing, 1963.

Rendel Harris, J. *The Cult of the Heavenly Twins.* Cambridge, U.K.: Cambridge University Press, 1906.

Rees, Alwyn, and Brinley Rees. *Celtic Heritage.* London: Thames and Hudson, 1978.

Reichel-Dolmatoff, Gerardo. *The Forest Within.* Dartington, U.K.: Themis Books, 1996.

Rhys, J. *Celtic Folklore.* Oxford: Clarendon, 1901.

———. *Lectures on the Origins and Growth of Religion.* London: Longmans, Green, 1892.

Runciman, S. *The Medieval Manichees.* Cambridge, U.K.: Cambridge University Press, 1947.

Rundle Clark, R. T. *Myth and Symbol in Ancient Egypt.* London: Thames and Hudson, 1959.

Santillana, Giorgio de, and Hertha von Dechend. *Hamlet's Mill*. London: Macmillan, 1970.

Seznec, J. *The Survival of the Pagan Gods*. Princeton, N.J.: Princeton University Press, 1972.

Skinner, Stephen. *Terrestrial Astrology*. London: Routledge and Kegan Paul, 1980.

Spence, Lewis. *The Mysteries of Britain*. Wellingborough, U.K.: Aquarian Press, 1979.

Stone, Merlin. *The Paradise Papers*. London: Virago, 1976.

Stutely, Margaret, and James Stutely. *A Dictionary of Hinduism*. London: Routledge and Kegan Paul, 1977.

Sullivan, William. *The Secret of the Incas*. New York: Three Rivers, 1996.

Tompkins, P. *Secrets of the Great Pyramid*. London: Penguin, 1978.

Tsunoda, T., et al. *Sources of Japanese Tradition*. New York: Columbia University Press, 1964.

Van Over, R. *A Chinese Anthology*. London: Pan, 1973.

Vermes, G. *The Complete Dead Sea Scrolls in English*. London: Penguin, 1998.

Vernant, J., and M. Detienne. *Cunning Intelligence in Greek Culture and Society*. Hassock, U.K.: Harvester, 1978.

von Franz, Marie-Louise. *Number and Time*. Trans. by Andrea Dykes. London: Rider, 1974.

Walker, A., trans. *Apocryphal Gospels, Acts and Revelations*. Edinburgh: T. and T. Clark, 1870.

Wallis Budge, E. A. *The Book of the Dead*. London: Arkana, 1989.

Werner, E. T. C. *Dictionary of Chinese Mythology*. Singapore: Kelly and Walsh, 1932.

Williams, C. A. S. *Encyclopaedia of Chinese Symbolism and Art Motives*. New York: Julian Press, 1960.

Williamson, R. W. *Religions and Cosmic Beliefs of Central Polynesia*. Cambridge, U.K.: Cambridge University Press, 1933.

Wilson, R. M. *Gospel of Philip*. New York: Harper and Row, 1962.

Wood Martin, W. G. *Traces of the Elder Faiths of Ireland*. London: Longmans, Green, 1902.

Zahan, Dominique. *Religion, Spirituality and Thought of Traditional Africa*. Trans. by Kate E. Martin and Lawrence M. Martin. Chicago: University of Chicago Press, 1983.

Zaehner, R. C. *The Dawn and Twilight of Zoroastrianism*. London: Weidenfield and Nicholson, 1961.

———. *Hindu Scriptures*. London: Everymans Library, 1968.

INDEX

Page numbers in **bold** represent figures.

BOOKS OF RELATED INTEREST

The Dimensions of Paradise
Sacred Geometry, Ancient Science, and the
Heavenly Order on Earth
by John Michell

Sacred Number and the Origins of Civilization
The Unfolding of History through the Mystery of Number
by Richard Heath

Matrix of Creation
Sacred Geometry in the Realm of the Planets
by Richard Heath

Temple of the Cosmos
The Ancient Egyptian Experience of the Sacred
by Jeremy Naydler

The Dream Culture of the Neanderthals
Guardians of the Ancient Wisdom
by Stan Gooch

The Genesis and Geometry of the Labyrinth
Architecture, Hidden Language, Myths, and Rituals
by Patrick Conty

Harmonies of Heaven and Earth
Mysticism in Music from Antiquity to the Avant-Garde
by Joscelyn Godwin

Cosmic Music
Musical Keys to the Interpretation of Reality
Edited by Joscelyn Godwin

INNER TRADITIONS • BEAR & COMPANY
P.O. Box 388
Rochester, VT 05767
1-800-246-8648
www.InnerTraditions.com

Or contact your local bookseller